Pulse Code Modulation
Systems Design

For a complete listing of the *Artech House Telecommunications Library*,
turn to the back of this book.

Pulse Code Modulation
Systems Design

William N. Waggener

Artech House
Boston • London

Library of Congress Cataloging-in-Publication Data
Waggener, William N.
    Pulse code modulation systems design / William N. Waggener.
      p.   cm.
    Includes bibliographical references and index.
    ISBN 0-89006-776-7 (alk. paper)
    1. Digital communications.   2. Pulse code modulation.   I. Title.
TK5103.7.W338   1998
621.382—dc21   98-41590
                                                                             CIP

British Library Cataloguing in Publication Data
Waggener, William N.
    Pulse code modulation systems design
      1. Pulse-code modulation—Design
    I. Title
    621.3'815365

    ISBN 0-89006-776-7

Cover design by Lynda Fishbourne

© 1999 ARTECH HOUSE INC.
685 Canton Street
Norwood, MA 02062   .

International Standard Book Number: 0-89006-776-7
Library of Congress Catalog Card Number: 98-41590

10 9 8 7 6 5 4 3 2 1

For my father, who took the time to show me the world from his shoulders in small towns across the United States.

Contents

	Preface	**xi**
	Acknowledgments	**xv**
1	**Pulse Code Modulation Systems**	**1**
1.1	System Models	1
1.1.1	Simple Transmission Model	5
1.1.2	Transmission Model With a Transponder	7
1.2	System Parameters	8
1.2.1	Signal-to-Noise Ratio	9
1.2.2	Bit Error Rate	11
1.2.3	Root Mean Square (RMS) Error	12
1.2.4	Timing and Synchronization	13
1.3	System Design Worksheet	14
	Reference	17
2	**Sampling, Quantization and Noise**	**19**
2.1	Sampling	20
2.1.1	Nyquist Sampling	20
2.1.2	Aliasing	22
2.1.3	Sample Rate Reduction	26

2.2	Quantization and Analog-to-Digital Conversion	29
2.3	Noise and Interference	35
2.3.1	Noise	36
2.3.2	Interference	45
2.4	Design Charts	49
	References	50
3	**Wireless Systems**	**51**
3.1	Radio Systems	51
3.1.1	Radio Frequency (RF) Propagation	53
3.1.2	Antennas	87
3.1.3	Background Noise and Interference	96
3.2	Optical Systems	99
3.2.1	Propagation	99
3.2.2	Optical Components	103
3.2.3	Noise	107
3.3	Design Charts	109
	References	113
4	**Cable Systems**	**115**
4.1	Wireline Systems	118
4.1.1	Propagation	118
4.1.2	Cable Characteristics	125
4.1.3	Cable Drivers and Receivers	131
4.2	Fiber Optic Systems	131
4.2.1	Propagation	133
4.2.2	Couplers	138
4.2.3	Transmitters	139
4.2.4	Detectors	140
4.3	Design Charts	146
	References	149

5	**PCM Encoding and Modulation**	**151**
5.1	PCM Baseband Codes	152
5.1.1	Symbol Codes	153
5.1.2	Channel Codes	164
5.2	Modulation and Multiplexing	174
5.2.1	Amplitude Modulation	176
5.2.2	Frequency Modulation	180
5.2.3	Phase Modulation	182
	References	185
6	**Demodulation and Detection**	**187**
6.1	Detection Basics	188
6.1.1	Carrier Versus Baseband Signaling	194
6.2	Baseband PCM Detection	198
6.2.1	Rectangular Symbols	198
6.2.2	Nyquist Symbols	201
6.2.3	Partial Response Signals	205
6.3	Coherent Demodulation	210
6.3.1	Single Sideband Amplitude Modulation (SSB/AM)	211
6.3.2	Quadrature Amplitude Modulation (QAM)	211
6.3.3	PSK and M-ary PSK	215
6.3.4	Frequency Shift Keying (FSK) and Continuous Phase Modulation (CPM)	216
6.3.5	Spread Spectrum	221
6.3.6	Coherent Demodulation of Optical Signals	221
6.4	Incoherent Demodulation	226
6.4.1	Amplitude Modulation	227
6.4.2	Frequency Modulation	228
6.5	Bit Error Performance With Intersymbol Interference and Fading	234

6.5.1	Intersymbol Interference	235
6.5.2	Fading	240
6.6	Error Correction	242
6.6.1	Block Codes	246
6.6.2	Convolutional Codes	249
6.6.3	Concatenated Codes	256
6.6.4	Trellis Code Modulation (TCM)	259
6.7	Equalization	259
6.7.1	Filtering	260
6.7.2	Transverse Digital Equalizers	261
6.7.3	Maximum Likelihood Sequence Estimation (MLSE)	263
	References	265
7	**Synchronization**	**269**
7.1	Synchronization Basics	269
7.2	Carrier Synchronization	280
7.2.1	Carrier Synchronization for Suppressed Carrier Systems	282
7.2.2	Performance of PSK Synchronizers	287
7.3	Symbol Synchronization	296
7.3.1	Symbol Synchronizer Performance	299
7.4	Code Synchronization	302
7.4.1	Block Codes	303
7.4.2	Convolutional Codes	305
7.5	Format Synchronization	305
7.6	Equalizer Training	307
	References	308
	About the Author	**311**
	Index	**313**

Preface

Pulse code modulation (PCM) is the dominant communications technology for many applications including telecommunications, telemetry, data recording, and data acquisition. PCM holds a theoretical performance advantage over older analog systems and is ideally matched to semiconductor technology. This book is about PCM systems design, the art and practice of conceiving and specifying a PCM communications system to satisfy a set of general requirements. Systems design views a problem from a "big picture" standpoint, starting with the end-to-end requirements and defining each component of the system to meet the overall requirements. In one application, a user may wish to design a system to transmit data from a number of sensors onboard an aircraft to a ground station for display and processing. In another application, a number of voice and data channels are to be transmitted over a fiber optic cable from one city to another. In each of these applications, the systems designer must conceive the systems architecture to perform the functionality and determine the architecture and performance of each major component in the system to satisfy the overall requirements.

The systems engineer is typically involved in the beginning stages of a project starting with the conception, the proposal, and the preliminary design. In these stages, breadth of knowledge is generally more important than depth. The systems designer needs to know 80% of many things and, ideally, 90% of a few key technologies. This book is directed toward this audience and attempts to provide the 90% of what is necessary to do a preliminary design of a communication system. The remaining 10% of the design must be provided by specialists in the key components of the system.

Over the course of 30 plus years of designing and analyzing a variety of PCM systems, I have nearly worn out some classic references while other texts and handbooks gather dust on my bookshelves. I have found myself collecting essential information in notebooks to provide a central reference. This book reflects that notebook, gathering up well known facts and data which I have found useful in the early design stages of a PCM system.

Audience

Audiences for this book include practicing communication systems engineers, electrical engineers which are working with pulse code modulation, and electrical engineering students majoring in data communications, data recording, command and control, or telemetry.

Contents

This book focuses on the communication system design starting with signals at one end of the system and sending them, using PCM technology, to the receiving end. The emphasis is on applied engineering and the author likes to think of this book as a notebook to be used both as a guide for communications system design as well as a desktop reference. For example, the bit error performance of the majority of PCM signaling methods is discussed and charts are provided for future reference. PCM is used in many diverse applications and transmission losses for cable (both wireline and fiber optic) and wireless media is considered with methods for quickly estimating system performance.

Chapters 1 and 2 introduce some basic concepts for PCM systems and define fundamental performance parameters. Chapters 3 and 4 consider the characteristics of the transmission channel. The overall performance of the PCM system is normally determined by the transmission channel and methods for estimating transmission loss are presented. Chapters 5, 6 and 7 get to the heart of the PCM system discussing PCM symbols, modulation, demodulation, detection, and synchronization.

Using This Book

As a notebook for practicing systems engineers, it is intended both as a tutorial source and as a desktop reference. It may be used as a text for a short course on

PCM systems. A three-day course might, for example, cover Chapters 1 through 3 on the first day, Chapters 4 and 5 on the second day and Chapters 6 and 7 on the third day. As a working reference, this book can answer many of the day-to-day questions which may arise in the preliminary design of a PCM system. During the detailed design phase, the book can serve as an index to more specialized texts. Chapters 5 through 7 can be supplemented by *Pulse Code Modulation Techniques* by Waggener, *Digital Communications* by Messerschmitt, and *Detection, Estimation and Modulation Theory* by Van Trees. Additional references are provided at the end of each chapter.

References

[1]   Waggener, Bill, *Pulse Code Modulation Techniques*, New York: Van Nostrand Reinhold, 1995.

[2]   Lee, E. A., and D. G. Messerschmitt, *Digital Communications*, Second Ed., Boston, Kluwer Academic Publishers, 1994.

[3]   Van Trees, Harry L., *Detection, Estimation and Modulation*, 3 volumes, New York: John Wiley and Sons, 1968.

Acknowledgments

Many individuals, past and present, have influenced my work, personally and professionally. Herman Moench and Bob Strum at Rose Hulman Institute of Technology taught me to think like an "electron," a dying art. Most recently, Sven Ridder, Lars Bergstrom, and the boys at B&R provided an exciting environment for new technology. Lars, we miss you. My wife Kathy, as always, provides constant support and contributed to much of Chapter 6.

The staff of Chestnut Mountain Group is an ongoing source of support. Bill Waggener, Jr. provides a forum for discussing the latest in biomedical technology while Genevieve Waggener supplies linguistic and acoustic technology. Lisa and Rick Roth support our education technology ventures and provide software support. Carol Waggener, although on sabbatical leave to Salt Lake City, continues to keep us on our toes. The CMG associates, Mary Roth, Libby Roth, Andrew Roth, Ariane Beauchamp and Daphne Waggener have always been willing to assist me at the computer while writing the book. Thanks to the anonymous reviewer and the staff at Artech for all of your help.

Bill Waggener
Sarasota, FL
Rabun Gap, GA
November 1998

1

Pulse Code Modulation Systems

Pulse code modulation (PCM) is the technology of transmitting information from one point to another by means of a pulse train which may be modulated in amplitude, frequency, or phase. In the most common realization, binary information is conveyed by a series of pulses which can only take one of two possible forms. The design of a system using this technology must consider the encoding of the original signal into digital form, the transmission (also referred to as the channel) media, the type of PCM used for transmission, the detection and decoding of the channel signal, and the reconstruction of the signal into the desired form at the receiving end.

PCM technology has been applied to many applications. The systems can be categorized in a number of ways: by the transmission media, by communication mode (one way or duplex), by system application (telecommunications, telemetry), and so forth. Despite the differences in applications, the PCM systems share a common model at the communication media layer.

1.1 System Models

PCM systems are divided into two generic types, a system with one information source communicating to one information recipient (termed a "sink") and a system with multiple sources and multiple sinks as illustrated in Figure 1.1. The first system is called a "point-to-point" system while the second type is called a "networked system." A general model of the point-to-point system is shown in Figure 1.2. The output of an information source, such as a sensor or a telephone,

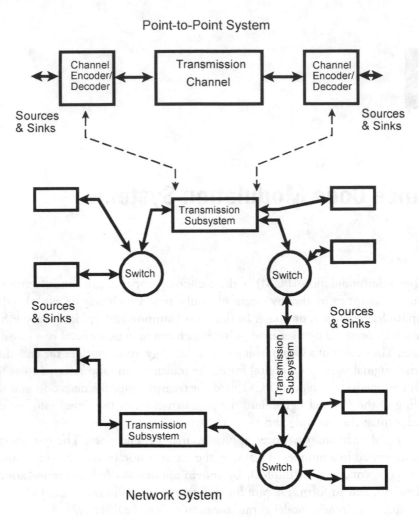

Figure 1.1 Generic PCM system models.

is fed to a source encoder which converts the source input to digital form and, possibly, performs a transformation of the digital information. Once the source has been encoded, the binary data stream is fed to a channel encoder. The channel encoder converts the binary data into symbols appropriate to the transmission channel. Channel encoding may also include the generation of an error correcting or detecting code. The channel encoder is followed by a modulator which matches the symbol stream to the characteristics of the transmission channel.

After transmission, the received signal is demodulated, the symbols are decoded and the binary data stream is recovered from the symbols. At this point

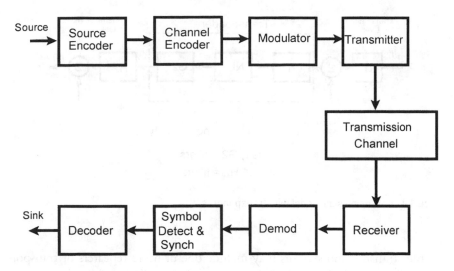

Figure 1.2 Point-to-point system model.

the binary data may be converted back to its original form (analog) or it may be used in digital form. At the receiving end there are many things going on behind the scenes. Before the signal can be demodulated, it may be necessary to extract a carrier reference to be used in the demodulator. Before the symbols can be recovered, a symbol timing signal must be extracted.

The key to PCM system performance is the characteristic of the transmission channel. The transmission channel (also referred to as the "communications link" or just the "link") can be as simple as an equivalent time-varying linear filter with an additive noise source as shown in Figure 1.3. This model is adequate for many PCM systems including wireless and cable systems. More

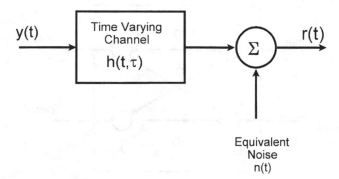

Figure 1.3 A simple transmission channel model.

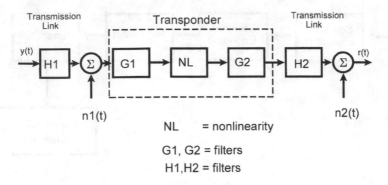

Figure 1.4 A transmission model with transponders.

complex transmission systems may include one or more repeaters or transponders as illustrated in Figure 1.4. Communication satellites, cable television systems, and long distance telecommunication systems all fit the more complex model. The nonlinear component of the repeater model greatly complicates the analysis of the transmission channel requiring, in many cases, computer simulation for accurate analysis.

The goal of the systems analyst should be to reduce the effects of the real transmission channel to a simple binary, symmetric channel as shown in Figure 1.5. In this model, binary "ones" and "zeros" are transmitted correctly from the source to sink with probability $(1 - p)$ and incorrectly with probability p. The same model is applied to both the point-to-point and the networked systems. There are occasions in which the channel is not symmetric, (i.e., the probability of error is different for "ones" than for "zeros") but, in this book, only symmetric channels are considered.

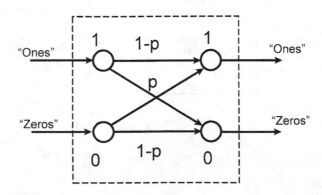

Figure 1.5 The binary symmetric channel model.

1.1.1 Simple Transmission Model

Figure 1.6 shows a simple transmission model including the modulator/demodulator (if required), the transmitter and receiver, couplers to the transmission media, and a linear, time-varying model of the transmission media. A single, additive noise source at the input to the receiver represents both receiver noise and noise introduced in the transmission channel. Each component in this model will be described in subsequent sections.

1.1.1.1 Modulator

The frequency spectrum of the symbols from the channel encoder normally extends to very low frequencies, including dc for some symbols. Signals with significant energy at low frequencies will be termed "baseband" signals. Many transmission channels do not support very low frequency signals and require the signals to occupy a band of frequencies centered about a "carrier" frequency. To accommodate these baseband signals, a modulator is required to translate the baseband signal spectrum to a bandpass spectrum. In the transmission model, the output frequency spectrum of the modulator is an important system characteristic. In radio systems, there are federal and international regulations which limit the power level outside of the nominal allocated bandwidth for various

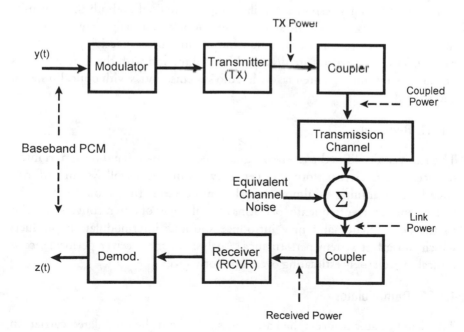

Figure 1.6 Complete transmission subsystem model.

types of systems. The PCM system engineer must be aware of these limitations in selecting the modulation and maximum symbol rate.

1.1.1.2 Transmitter

The transmitter provides signal power gain to overcome transmission channel losses. Maximum power gain is often provided with nonlinear amplification. The nonlinearities introduce harmonic distortion in the signal, creating spurious signals which may violate transmission channel requirements. At the transmitter end these spurious signals may be eliminated by additional filtering. The problem with spurious signals is most acute in systems with repeaters or transponders and at the receiving end where the spurious signals may be strong enough to cause transmission errors.

1.1.1.3 Coupler

For want of a better word, coupler is used to describe the generic function which couples the transmitter signal into the transmission media. In a radio system, this function is provided by antennas. In optical systems, this function is provided by optical lenses or fiber optic pipes. In cable systems, this function is provided by transformers and electrical matching networks. Whatever the application, the coupler is characterized by a transfer function with gain (or loss) and a frequency response. Similar couplers are used on both the transmitting and receiving ends of the system. The couplers normally obey reciprocity, that is, they work equally well as a transmitting unit or as a receiving unit. Typical coupler characteristics are discussed in conjunction with the different types of transmission media, antennas with radio systems, optics with optical systems, and so forth.

1.1.1.4 Receiver

The receiver amplifies the received signal rejecting out-of-band noise and interference. The receiver performance generally dominates overall system performance by determining minimum signal-to-noise ratio. In a modulated carrier system, the receiver frequently performs several stages of down conversion prior to the demodulator and may introduce spurious intermodulation products which can affect system performance. Analysis of the receiver performance is critical to accurately estimating overall performance.

1.1.1.5 Demodulator

The demodulator recovers the baseband signal from the modulated carrier. In coherent systems, the demodulator must recover an estimate of the carrier reference signal from the noisy received signal. Carrier recovery is an important aspect of many PCM systems and is covered in greater detail in Chapter 7.

1.1.2 A Transmission Model With a Transponder

In many large systems, transmission losses are such that the signal needs to be periodically regenerated to overcome the losses. One of the best examples of such a system is the "bent pipe" communications satellite as shown in Figure 1.7. A PCM multiplex signal is transmitted from a ground station to an earth orbiting satellite. The signal is received at the satellite, amplified and retransmitted back to another ground station. Simplicity is a major advantage since the PCM baseband signal does not need to be recovered onboard the satellite. A disadvantage of this system is that noise and distortion are added to the signal in a nonlinear fashion at the satellite, degrading overall performance.

In other examples, the transponder may be called a repeater as is the case in long-line telecommunication systems. In these systems, many repeaters are required for long distance transmission. The number of repeaters required generally precludes the use of simple transponders because of the cumulative effect of the noise and distortion introduced at each stage. Instead, the baseband signal is regenerated at each repeater. In this case, the overall link can be modeled as a cascade of binary symmetric channels. There are, however, some cumulative effects which may not be obvious at first glance. At each repeater a symbol clock must be reconstructed and some jitter is introduced in the clock by signal noise, distortion, and spectral characteristics. This jitter is propagated to the next repeater where additional jitter may be added. In systems with many repeaters,

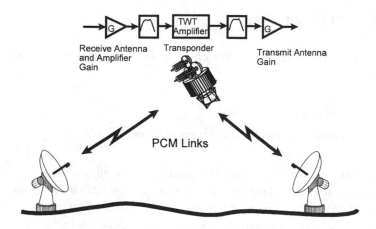

Figure 1.7 A "bent pipe" satellite link.

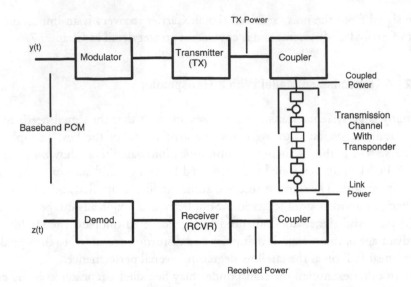

Figure 1.8 A transmission subsystem model with a transponder.

performance may be limited by timing jitter and not by the cascading of bit errors.

A simplistic model of a transmission channel with a transponder is shown in Figure 1.8. Even this simple model may be difficult to analyze without the use of computer simulation. Consideration of this model will be provided in the discussion of wireless radio links.

1.2 System Parameters

The PCM systems engineer is concerned with two aspects of the system performance:

- System error performance;
- The time required for the system to become stable.

How the error performance is specified and used depends on the application. In some systems the user wants to know the expected bit error rate from end-to-end. In other systems, the number of "errored seconds" in a given period is required, while in other systems, the end-to-end accuracy is important. Regardless of the performance criteria, the effective bit error rate is of prime concern.

The time required for the system to become stable is very important but often overlooked in system design studies. In a PCM system, a finite period of time is required for the demodulators, symbol detectors, and decoders to achieve "lock." These times can be particularly significant in packet systems and systems which require transmission acknowledgment. Both the error rate and synchronization times can affect the overall system response. There are many examples of systems where the overall system transmission rate is determined by these factors.

1.2.1 Signal-to-Noise Ratio

Intuitively, system performance is determined by the signal-to-noise ratio in the transmission links in the system. For any given transmission link, a graphical depiction of the signal and noise power levels at various points in the system can be very useful in understanding and analyzing the performance. An example of a signal power level diagram is shown in Figure 1.9 for a radio system. In this example, the transmitter power level is increased by the antenna gain. The transmission link median loss is shown followed by the receiving antenna gain. The received signal power level is shown on the right-hand axis. The equivalent input noise level is also shown on this axis from which the median signal-to-noise power ratio is determined. The radio frequency path loss is variable as indicated by a probability distribution function shown along the axis. If the

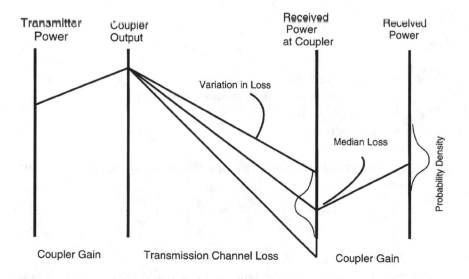

Figure 1.9 Signal power level diagram for a radio system.

distribution indicates the probability that the loss will be a certain value, then the probability that the signal-to-noise ratio will be less than a given value can be estimated.

The definition of signal-to-noise ratio will depend to some extent on the application but it will normally be defined as the ratio of the root mean square signal power to the root mean square noise power with the equivalent noise bandwidth implicitly defined.

$$SNR \doteq \frac{P_{signal}}{P_{noise}} \qquad (1.1)$$

The noise is characterized by the power in a small frequency band as a function of frequency. This is known as the "noise spectral density," $N_0(f)$. The total noise power is obtained by integrating the noise spectral density over the receiver bandwidth. When the noise has a constant spectral density, it is commonly called "white" noise and the noise power in a given bandwidth is the product of the spectral density and the equivalent noise bandwidth, $N_0 B_N$.

An alternative definition for signal-to-noise ratio which will be useful for many PCM systems with white noise is the energy contrast ratio

$$\frac{E_b}{N_0} \doteq \frac{\int_0^T s^2(t)dt}{N_0} \qquad (1.2)$$

where

$s(t)$ = the signal waveform
N_0 = the noise spectral density (assumed constant)
E_b = the energy per bit
T = the bit period

For the case of a rectangular pulse with amplitude, A, the energy contrast ratio is

$$\frac{E_b}{N_0} = \frac{A^2 T}{N_0} = \frac{A^2}{N_0 \frac{1}{T}} \qquad (1.3)$$

For this case, the energy contrast ratio is the signal-to-noise ratio in a noise bandwidth equal to the bit rate.

In more general terms, the energy contrast ratio may be defined in terms of the symbol energy rather than the bit energy. When one bit of information is conveyed by one symbol, the symbol energy is equivalent to the bit energy. When more than one bit is conveyed by the symbol (a multilevel symbol, for example), the energy per bit is related to the symbol energy by

$$E_b = \frac{E_s}{m} \qquad (1.4)$$

where

E_s = the energy per symbol
m = the number of bits per symbol

The characterization of noise and the computation of noise power in PCM systems is considered in greater detail in Chapter 2.

1.2.2 Bit Error Rate

The error rate of a PCM system (whether bit errors or symbol errors) is monotonically related to the signal-to-noise ratio. This can be demonstrated by the following argument. Suppose a PCM system uses a channel symbol which conveys one bit by taking on one of two possible levels. The received signal with additive noise is illustrated in Figure 1.10. The noise is assumed to have a probability density which causes the received signal to have a distribution as shown. The receiver must decide which level was transmitted and uses a decision threshold to decide on one level if the threshold is exceeded and the second level if the threshold is not exceeded. The probability of making an error is then proportional to the area under the noise probability density function as illustrated. The integral of the density function is the probability distribution function which can be shown [1] to be monotonic. As the signal-to-noise increases, the area under the density curve starting at the threshold level decreases and, hence, the error probability decreases.

The error rate in a system can vary just as other parameters. In many applications, the average error rate is an acceptable measure of performance. For some systems, however, average error rate can be very misleading. For example, some systems are prone to large, but infrequent, bursts of errors which may have a low average error rate but unacceptable performance. In such systems the designer needs to concentrate on the conditions which create error bursts and attempt to combat these errors.

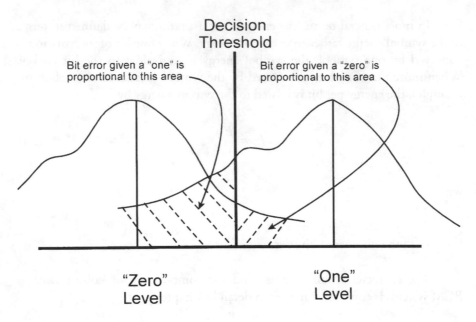

Figure 1.10 Received signal with noise density.

1.2.3 Root Mean Square (RMS) Error

One measure of the performance of a system is the end-to-end error between the input signal(s) and the corresponding output signal(s). There are four possible combinations of input and output signal formats:

- Analog input, analog output;
- Analog input, digital output;
- Digital input, analog output;
- Digital input, digital output.

The telephone system is a good example of the first case in which the speaker's voice signal is digitally encoded, transmitted by PCM and reconstructed to an analog signal in the receiver's handset. The second case is typical of a telemetry system: encoding analog signals from sensors for transmission to a computer system. A computer controlled supervisory control system is an example of the third type of system with digital control signals transmitted via PCM to analog controlled devices. Finally, the remaining type is typical of computer-to-computer data communications. The technology trend is increasingly toward the latter type of system with analog portions shrinking toward the extreme ends of the system. In all of these systems there is an interest in expressing some measure of the error between input and output.

Starting with an analog-to-analog system, errors are introduced in the sampling and analog-to-digital conversion, bit errors in the transmission process, and errors in the digital-to-analog process. The second type of system is similar to the first except for the digital-to-analog conversion errors. The third type has only the transmission bit errors and the reconstruction errors while the fourth system has only the errors due to transmission.

If we look at the effect of transmission errors on the error in a data sample and assume bit errors are equally likely in any bit within a word, the probability of no errors in a word with a length of L bits is

$$P_n = (1 - p)^L \approx 1 - Lp \tag{1.5}$$

The probability of at least one error is

$$P_e = 1 - P_n \approx Lp \tag{1.6}$$

For random errors, if an error occurs it is most likely a single bit error and a one half scale error would be the largest error. Unlike other sources of error, it is not reasonable to associate a root mean square error with an isolated word error unless the bit error rate is exceptionally high. For example, a system with a 10 bit word and a bit error rate of 10^{-5} would only experience a word error about once in 10,000 words. Even with a 10^{-3} error rate, only about one word in one hundred would be expected to be in error. For these examples it is more reasonable to treat the errors as "outliers," samples which would be discarded if they are significantly larger or smaller than their surrounding samples.

1.2.4 Timing and Synchronization

Transmission errors are not the only important degradation associated with PCM systems. It takes a finite period of time to achieve synchronization in a PCM system and, in some cases, timing jitter can introduce system errors. Serious system problems have been created by a failure to consider the synchronization of the PCM link. The time to acquire synchronization is inversely proportional to the synchronizer bandwidth while the ability to achieve synchronization in a noisy environment is dependent on a narrow bandwidth. Thus, if the preamble of a data packet is not long enough, the synchronizer may be unable to use a narrow enough bandwidth to lock on to a noisy signal. These issues will be addressed in Chapter 7.

1.3　System Design Worksheet

The starting point for an analysis of a PCM system is the transmission link. In the case of a complex system with multiple links, each link needs to be separately analyzed. As indicated previously, the goal is to reduce all transmission links to equivalent binary symmetric channels. The approach of the systems engineer should be to start with pen, paper, and calculator to perform an initial assessment of the link performance. This analysis should then be successively improved using more powerful methods, as appropriate. No matter what the transmission channel, the first place to start is a worksheet listing the signal and noise levels at various points within the system. The worksheet is ideally suited for implementation as a spreadsheet but a pencil and paper implementation also works. The worksheet is customized to the application but a generic template can be used as a starting point. The point-to-point transmission channel will be the basis for the generic worksheet. The worksheet tabulates the transmitter power, the coupler gain (or loss), the channel propagation loss, the receiver coupler gain, and the receiver noise. From the basic worksheet, the nominal signal-to-noise ratio can be estimated. Using charts for the bit error performance for the type of signaling used, the error rate can be estimated. This initial analysis can be refined by including a more detailed analysis of the propagation losses with consideration of statistical variations in the loss.

The generic worksheet is shown in Table 1.1. Gains and losses in a system are multiplicative and it is traditional to express these factors in decibels (dB). The worksheet provides a column for listing parameter values for reference, but the computations are all performed in dB. Power levels are normally of interest so that the calculation of dB from a value is $10 \, log_{10} \, (V)$, where V is the signal power. Strictly speaking, the dB calculation is nondimensional; however, unit modifiers are often attached to clarify the parameter. For example, dBm refers to a power level expressed in milliwatts (mW).

The generic worksheet provides entries at the top of the sheet to record general parameters of the system, items such as the transmission bit rate, the type of modulation used, and so forth. For the point-to-point transmission link, the sheet begins at the transmitter (abbreviated TX) power output expressed as watts (or milliwatts). Line 2 of the sheet is the coupler gain (in the case of a radio link, the transmitting antenna gain) and line 3 computes the effective power as the sum of lines 1 and 2. The transmission loss is

Table 1.1
Generic Worksheet

Project #		System Worksheet		Date:	
System parameters					
Transmission carrier frequency		MHz	Modulation		
Bit rate		Bits per second	Transmission link distance		km
Line #	Parameter	Value	Units	dB	Remarks
1	TX power		Milliwatts		dBm
2	TX coupler gain				dB
3	Effective power		Milliwatts		dBm
4	Transmission loss				dB
5	RCVR coupler gain				dB
6	Received power		Milliwatts		dBm
7	RCVR noise figure				dB
8	Noise spectral density		Milliwatts per Hz		dB/Hz
9	RCVR noise bandwidth		Hz		dBHz
10	Noise power		Milliwatts		dBm
11	SNR				dB
12	Required SNR				dB
13	Design margin				dB

recorded on line 4 with the receiver (abbreviated RCVR) coupler gain on line 5. The received power in dBw, or dBm, is the sum of lines 3, 4, and 5, with gains as positive values and losses as negative values. With the received power computed, the receiver noise power must be estimated. The receiver noise power may be known from measurements or published specifications, in which case, the power is entered on line 10. In some cases, the receiver noise performance is expressed as a "noise figure," a factor relating the noise to the theoretical minimum noise. The noise power (line 10) is the product (or sum of dBs) of the noise figure, the theoretical noise spectral density and the receiver noise bandwidth, the sum of lines 7, 8, and 9. The received signal-to-noise ratio (SNR) is the ratio of received signal power to noise power

Table 1.2
A PCM Data Link Example

Project # Bill's Foley			System Worksheet		Date: 8/20/07	
System parameters			Modulation	PSK		
Transmission carrier frequency	2200		MHz			
Bit rate	10^7		Bits per second	Transmission link distance	32	km
Line #	Parameter	Value	Units	dB		Remarks
1	TX power	5000	Milliwatts	37		dBm
2	TX coupler gain			−2		dB
3	Effective power		Milliwatts	35		dBm
4	Transmission loss			−140		dB
5	RCVR coupler gain			24		dB
6	Received power		Milliwatts	−81		dBm
7	RCVR noise figure			10		dB
8	Noise spectral density		Milliwatts per Hz	−174		dB/Hz
9	RCVR noise bandwidth		Hz	70		dBHz
10	Noise power		Milliwatts	−94		dBm
11	SNR			13		dB
12	Required SNR			12		dB
13	Design margin			1		dB

or the sum of lines 6 and 10. Finally, from consideration of the type of signal transmitted and the required system performance, a minimum required SNR is determined and entered on line 12. Subtracting line 12 from line 11 gives the system design margin.

The generic worksheet is greatly simplified and the systems engineer should customize the worksheet according to the application. As the types of systems are discussed, the worksheet can be refined according to the application. Implementing the worksheet on a spreadsheet program is a trivial task. Spreadsheet programs are available for all size computers including handheld units so that worksheets can be an essential and portable tool.

Example 1.1 A PCM Data Link

A PCM data link between a ground station and an unmanned airborne vehicle (UAV) is required to transmit a serial data stream at a rate of 10 megabits per second. A phase shift keyed (PSK) modulation on a 2200 MHz carrier is proposed with a dipole antenna on the UAV and a 6 ft tracking antenna on the ground. At the maximum range of 20 miles, the transmission loss is estimated to be 140 dB. A 5W transmitter is used on the UAV and the ground receiver has a 10 dB noise figure. From these parameters, the worksheet is created as shown in Table 1.2. The transmitter power in dBm is 37 dBm (values will be rounded to nearest whole number) and the estimated UAV antenna gain, including other losses, is −2 dB. The effective radiated power is 35 dBm. With the transmission loss and the receiver antenna gain, the received power is estimated to be −81 dBm. Using the receiver noise figure, the calculated noise spectral density (a constant value, to be discussed in Chapter 2), and an assumed receiver noise bandwidth equal to the bit rate, the receiver noise power is estimated to be −94 dBm. The estimated SNR is computed to be 13 dB. From a consideration of the required performance, a 12 dB SNR is required. For the parameters given and the assumptions made, the data link design margin is only 1 dB. This quick analysis shows that design assumptions and parameters must be carefully refined to determine if the margin estimate is accurate. The system losses are statistically variable and the design margin can be woefully inadequate in a reliable system.

Reference

[1] Papoulis, A., *Probability, Random Variables, and Stochastic Processes*, New York: McGraw-Hill Book Co., 1965.

2

Sampling, Quantization, and Noise

Starting at the source, the source encoder is the first major subsystem. The functions of the source encoder depend on the application. If the source is an analog signal, the source encoder includes a sampling and a quantization function. The quantization function is normally termed the analog-to-digital convertor, or ADC. Additional digital processing may be done on the digitized samples before formatting the samples into the PCM format. Source compression is a common digital processing function applied to the signal samples.

With increasing frequency, the input to the PCM system is already in digital form. This is true of computer-to-computer links; many telecommunication systems convert analog signals into digital words using single integrated circuits at the source. With these systems, the only concern of the systems engineer is the sensitivity of the digital words to system errors. If the samples are compressed, system errors can seriously degrade overall performance. On the other hand, if the samples were highly over sampled (for example, nearly dc signals) the system may be tolerant to system errors. Though the systems engineer may have no control over the source encoding, it is important to understand (if possible) the total application to optimize the PCM design.

Just as sampling and quantization are basic functions in the PCM system, an understanding of noise and interference is key to overall performance. To compute system performance, noise and interference must be modeled and characterized. Noise is generally considered a statistically random component characterized by a probability density and a power spectrum (or autocorrelation). Interference components are quasi-deterministic such as power supply pickup or intermodulation signals. The effect on performance of both random noise and interference must be considered.

2.1 Sampling

One axiom of digital systems taught to engineers is the sampling theorem. The theorem states that the sampling rate in a digital system must be at least twice the highest signal frequency; the minimum sampling rate is normally called the Nyquist rate. While the principle is simply stated, the theorem is often misused. Common mistakes include sampling at twice the highest frequency of interest without bandlimiting the signal bandwidth to that frequency. You may only be interested in frequencies up to 2 kHz but the signal may contain frequency components well above that frequency which must be filtered out before sampling. Another common mistake is to take the sampling theorem literally and sample at exactly the Nyquist rate without considering the system performance implications. Finally, there is the question of the required sampling rate of modulated signals. If a 1 kHz sinewave modulates a 100 kHz carrier, do you have to sample at 200 kHz? Intuitively the answer is no, a sampling rate of two kilosamples per second should be adequate.

2.1.1 Nyquist Sampling

A simple model of the sampling process is illustrated in Figure 2.1. The sampler is represented by an ideal multiplier that multiplies the input signal by a sam-

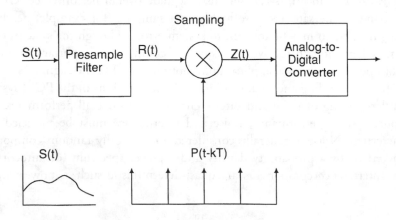

Figure 2.1 Model of sampling process.

pling function represented by an infinite series of impulses. This is essentially an amplitude modulator with a pulse train for a carrier. Amplitude modulation produces carrier components with sidebands on either side of each carrier component starting at dc and proceeding up through components at multiples of the sampling rate.

From a mathematical standpoint, the sampler output is the product of the input signal and the sampling function

$$y(t) = r(t) \cdot p(t) \tag{2.1}$$

From Fourier transform theory [1], the frequency spectrum of the product of two functions is the convolution of the spectrum of the two signals

$$Y(\omega) = \int_{-\infty}^{+\infty} R(\omega - \lambda) \cdot P(\lambda) d\lambda$$

or as a shorthand notation

$$Y(\omega) = R(\omega) * P(\omega) \tag{2.2}$$

where * represents the convolution operation.

To convolve two functions, the overlapping area obtained by sliding one function past the other is computed. To illustrate, the signal frequency spectrum is assumed to be a simple, lowpass triangular function and the frequency spectrum of the impulse sampling function is also a train of impulses. Convolving these functions produces a frequency spectrum at the output of the sampler as shown in Figure 2.2.

The output spectrum of the sampler consists of a periodic signal repeating at multiples of the sampling rate. Even if the sampling function is not an ideal impulse train, any periodic sampling pulse train will exhibit similar behavior with components around all harmonics of the sampling rate. The illustration shows a case in which the highest frequency of the signal, f_s, is clearly less than one half the sampling rate, obeying Nyquist's theorem. The original signal can be recovered exactly if the sampled signal spectrum is filtered using an ideal lowpass filter with constant gain up to one half the sampling rate and zero response elsewhere. The key points here are that the input signal has no components, exceeding one half the sampling rate, and an ideal lowpass filter is used to

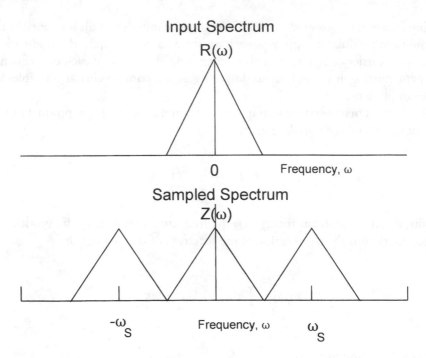

Figure 2.2 Sampled frequency spectrum.

reconstruct the signal. In practical systems, neither of these assumptions are true. In the real world, system errors are introduced in sampling and are termed "aliasing" errors. The goal of the system designer is to reduce these errors by careful choice of sampling rate, by presampling filtering, and by the design of the reconstruction filter.

2.1.2 Aliasing

When a signal has components exceeding the Nyquist frequency, the sampling process "folds" over components higher than one half the sampling rate. Figure 2.3 illustrates why this is termed "folding" for a signal with energy in the frequency band higher than the Nyquist rate. If the sampled signal is reconstructed with an ideal filter, the folded components represent an error called the "aliasing" error. The error components are called aliased components because they are higher frequencies masquerading as lower components. For example, a sinewave at a frequency of 0.75 the sampling rate will appear to be a component at 0.25 sampling rate. The aliasing error can be combated by three means: increas-

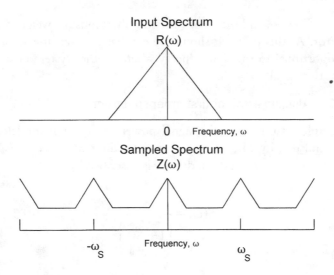

Figure 2.3 Sampled signal aliasing.

ing the sampling rate, bandlimiting the signal, and using a sharp cutoff reconstruction filter. In practice, a combination of all three methods is employed.

If the region around the folded spectrum is magnified, as in Figure 2.4, the aliased energy can be reduced by increasing the presampling filtering and by using a sharp cutoff reconstruction filter. The distortion of the signal created by

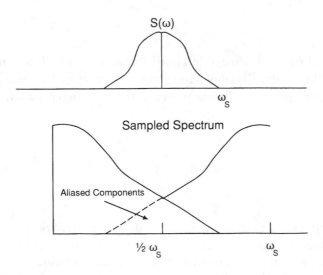

Figure 2.4 Aliasing error.

presampling filtering, and the reconstruction filter, must be weighed against the aliasing error. At this point, it should be emphasized that the aliasing error is always proportional to the signal energy whereas other system errors are independent of the signal.

Example 2.1 Aliasing error for first order spectrum

As an example, assume that the signal power spectrum is modeled as a flat power spectral density filtered by a first order filter with a 1 rad/s cutoff frequency. The signal power spectral density is described by

$$S(\omega) = \frac{1}{\left(1 + \dfrac{\omega^2}{\omega_0^2}\right)} \tag{2.3}$$

where

ω_0 = the -3 dB power frequency

The sampling rate will be normalized to the -3 dB frequency of one rad/s. The power aliased at half the sampling rate is

$$P_{alias} = \int_{\frac{\omega_s}{2}}^{\infty} S(\omega)\,d\omega \tag{2.4}$$

This integral can be evaluated in closed form and can be found in a table of integrals [2] or using a symbolic integration program such as Maple™ or Derive.™ Normalizing the aliasing power to the total signal power, the aliasing power is

$$P_{alias} = 1 - \frac{2}{\pi}\tan^{-1}\left(\frac{\omega_s}{2\omega_0}\right) \tag{2.5}$$

The aliasing error is normalized to the total signal power and plotted in Figure 2.5. To reduce the aliasing error to approximately 10% requires a sampling rate of about 12.5 times the -3 dB cutoff frequency and to reduce the error to 5% requires a normalized sampling rate of about 25. To reduce the aliasing error for

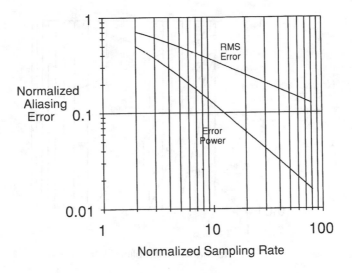

Figure 2.5 Total normalized aliasing power for a first order power spectrum.

a given sampling rate or, conversely, to reduce the required sampling rate for a given error, a higher order presampling filter (an "anti-aliasing" filter) is necessary.

The aliasing error is more frequently expressed as a root mean squared (RMS) value normalized to RMS signal value. The computed aliasing error for maximally flat power spectra with orders from one to five is shown in Figure 2.6. It should be emphasized that the aliasing error represents the total energy folded back from half the sampling rate without regard for any additional filtering.

Although the computed aliasing error is typically pessimistic, it does serve to illustrate the importance of presample filtering and choice of sampling rate to insure that aliasing error is minimized. Various applications have developed rules of thumb with respect to required sampling rate and presample filtering. In the telemetry and data acquisition fields, a common rule of thumb requires a sampling rate of at least 5:1 with third order presample filtering. In telecommunications, voice signals are sampled at 8 kilosamples per second. The power spectrum of voice typically has a peak energy between 800 and 1000 Hz, rolling off above the peak at 12 dB per octave. The 8 kilosamples per second sampling rate is thus approximately 10 times the peak frequency. In voice signals, perception of the errors is more important than some arbitrary percentage error and the sampling rate was chosen after many empirical studies.

The sampling of standard television video signals also must rely on empirical studies to assess the best sampling rate. If the sampling rate is too high, the PCM data rate is unacceptable, if the rate is too low, video quality suffers.

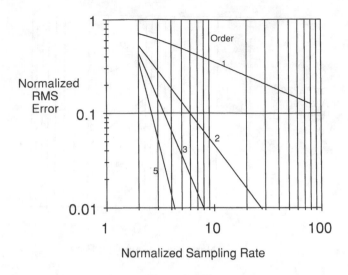

Figure 2.6 Aliasing error as a function of normalized sampling rate.

Direct sampling of the NSTC composite signal has been used for monochrome video using a sampling rate near the common IF frequency of 10.7 MHz. Using 8 bits per sample, the composite data rate is about 90 Mbps. Full color studio quality video encodes each color channel separately resulting in a raw data rate of about 270 Mbps. Recently the MPEG I and MPEG II video compression standards [3] have been developed reducing the data rate to less than 10 Mbps.

2.1.3   Sample Rate Reduction

The aliasing error can be minimized by simply increasing the sampling rate; however, this approach can increase system cost and degrade system perform-ance. It is later explained that system bit error performance depends on the en-ergy per bit so that shorter bit periods (higher rates) result in poorer error performance. There is a strong incentive to reduce the system sampling rate to as close to the Nyquist rate as possible. The advent of cost effective digital signal processing components makes this possible through a sample rate reduction technique. Designing presample analog filters with high stopband loss and a sharp transition between the passband and the stopband is both difficult and expensive to implement.

     Digital signal processing offers an alternative to the difficult analog pre-sample filter design problem. The basic idea is to use a lower order presample filter, sample at a high rate, digitally filter the sampled signal, and then reduce the sampling rate to near the Nyquist frequency. This process is illustrated in

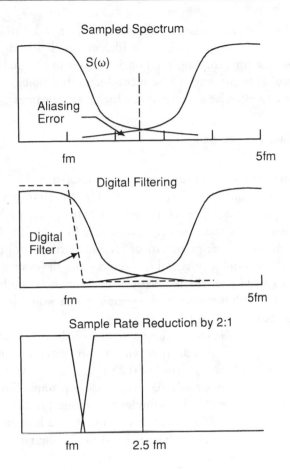

Figure 2.7 Subrate sampling concept.

Figure 2.7. In this example, an analog presample filter bandlimits the signal before sampling at five times the signal bandpass, f_m. A very sharp cutoff digital filter is used to bandlimit the signal to a bandwidth of 1.25 f_m. Because the signal is bandlimited to 1.25 f_m, the output sample rate of the digital filter can be reduced to 2.5 f_m. The sample rate reduction is termed "decimation." The aliasing power has not been eliminated by this method but is limited to the passband and the small amount of energy passed by the digital filter in the stopband.

In order to implement the subrate sampling method, the digital filter performs the required computations at the decimated rate. The processor must accept samples at the input sample rate but only needs to output the sample results at the decimated rate. The order of the digital filter required depends on the type of filter, infinite impulse response (IIR), or finite impulse response (FIR), and the filtering specifications. With current digital signal processing technology, the subrate sampling method is limited to a signal bandwidth of several hundred kHz, or less. As semiconductor technology improves, the method will be cost effective at increasingly higher sample rates.

Example 2.1 Subrate Sampling

Four sonar signals are to be transmitted over a line-of-sight PCM radio system. The highest signal frequency of interest is 30 kHz and 10 bit resolution (0.1% full scale) is required. Sampling each signal at 5 times the highest frequency (150 kilosamples per second) requires a data link bit rate of 6 Mbps without formatting overhead. At a sampling rate of 5:1 and a fifth order presample filter, the aliasing error for a full scale signal is about 5 times the desired 10 bit resolution. In order to achieve an aliasing error of the same magnitude as the 10 bit resolution with the fifth order filter, the sampling rate would have to be about 10:1, doubling the data rate to 12 Mbps.

Implementing a subrate sampling system to reduce the sampling rate to 2.5:1 (3 Mbps) requires a digital filter with a normalized transition band between the passband and the stopband of 0.25 f_m (7.5 kHz), with >30 dB stopband loss. A third order maximally flat amplitude presample filter will be used with a 7.5:1 sampling rate (225 kilosamples per second per channel). From the aliasing error chart, the RMS error at this sampling rate is about 1.8%. However, the aliasing error power in the passband of the digital filter is approximately

$$P_{alias} = \int\limits_{6.5f_m}^{\infty} \frac{1}{1 + \left(\dfrac{f}{f_m}\right)^6} \, df = 1.524 \cdot 10^{-5} \tag{2.6}$$

The aliasing error power in the passband can be expressed as a fraction of the total aliased power

$$P_{Normal} = \frac{P_{alias}}{\int\limits_{0}^{\infty} \dfrac{1}{1 + \left(\dfrac{f}{f_m}\right)^6} \, df} = \frac{1.524 \cdot 10^{-5}}{1.047} = 1.456 \cdot 10^{-5}$$

$$e_{rms} = \sqrt{P_{Normal}} = 0.0038 \qquad (2.7)$$

The square root of this ratio can be used to compute the relative aliasing error reduction by the subrate sampling. For a third order power spectra with a 7.5:1 sampling rate and a 3:1 decimation (2.5 f_m), the RMS aliasing error is about 0.38%. This does not meet the 0.1% goal but illustrates the method. A higher order presample filter is suggested to further reduce the aliasing error.

2.2 Quantization and Analog-to-Digital Conversion

Just as the sampling quantizes the signal in time, the analog-to-digital converter (ADC) quantizes the signal in amplitude. The signal samples are fed to the ADC which quantizes each sample into a number of discrete levels, L. The number of levels are then encoded into a binary word of m bits.

$$m = \log_2(L) \qquad (2.8)$$

A number of methods are used to implement the ADC process. Some of the more common methods are

- Flash;
- Half-flash;
- Voltage to frequency (VF);
- Single slope;
- Dual slope;
- Successive approximation;
- Differential encoding.

An inverse relationship exists between ADC resolution and sampling rate. The flash method converts the analog sample using parallel comparators and is the fastest but has the lowest resolution. The single slope, dual slope, and successive approximation ADCs are slower but have higher resolution. Figure 2.8 plots the resolution versus sampling rate for a number of commercially available ADCs and empirically shows the inverse relationship. As might be expected, the closer the ADC performance is to the state-of-the-art envelope, the more expensive the converter. The resolution/sampling rate plane is divided by the typical types of

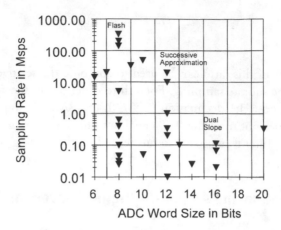

Figure 2.8 Analog-to-digital converter resolution.

applications. Video and radar systems typically require the 8 bit ADCs with sampling rates in the 100 to 500 MHz range. Telecommunication applications require 8 bit ADCs but with 8 kHz sampling rates. Telemetry and data acquisition systems use 10 to 16 bit ADCs with sampling rates in the tens to hundreds of kilohertz. Data loggers require high resolution ADCs (14 to 20+ bits) but can use low sampling rates.

In the PCM system, the signal is sampled both in time and in amplitude. The sampling in amplitude is called quantization. The resolution of an ADC is determined by the smallest input level distinguished by the encoder, ΔV. If the total range spanned by the encoder is L, the minimum resolved increment for an m-bit ADC is

$$\Delta V = \frac{L}{2^m} \tag{2.9}$$

Typically, the resolution of an ADC is expressed as the number of bits, m, of the ADC. The dynamic range of an ADC is the ratio of the largest level to the minimum increment, 2^m. A 10 bit ADC has a dynamic range of about 60 dB ($20 \log_{10} 1024$).

When a signal is quantized, there is an uncertainty of ΔV in the sample measurement. For a dynamic signal the instantaneous error will vary from sample to sample. The worst case error due to quantization will be $+/-$, one half the quantization step. If the input samples to the ADC are considered to be random variables, the quantization error can be assumed to be approximately equally likely within the quantization interval, ΔV. The probability density function for the error is

$$p(v) = \frac{1}{\Delta V}, \quad -\frac{\Delta V}{2} \leq v \leq \frac{\Delta V}{2} \quad (2.10)$$

The mean square quantization error is

$$\sigma_q^2 = \int_{-\infty}^{+\infty} v^2 p(v) dv = \frac{\Delta V^2}{12} \quad (2.11)$$

If the signal exceeds the full scale range of the ADC it is "clipped," and an overload error occurs. The overload can be viewed as a different form of quantization error or as harmonic distortion. Regardless of how one views this error, the error is serious and must be considered in the system design.

The overload error is zero for a sinusoidal input below the full scale range of the ADC. The signal-to-quantizing error ratio (SQR) for the sinewave as a function of the level of the signal relative to full scale is given by

$$SQR = 1.5\alpha^2 2^{2m}$$

where

α = signal level, relative to full scale
m = number of ADC bits

Expressed in dB

$$SQR_{dB} = 1.761 + 20 \, Log \, \alpha + 6m \text{ in } dB \quad (2.12)$$

If the sinewave exceeds the full scale range, the peak amplitude is clipped and the received signal consists of the original sinewave and its harmonics. The overload error, shown in Figure 2.9, is proportional to the area of the square of the clipped portion. The SQR for a sinewave as a function of the signal amplitude relative to full scale is shown in Figure 2.10 for a 10 and 12 bit ADC.

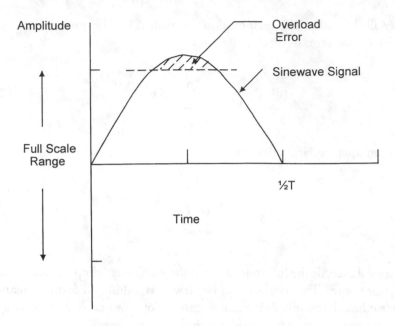

Figure 2.9 ADC overload error.

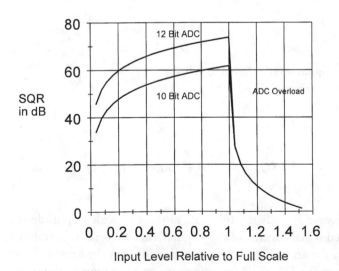

Figure 2.10 Signal-to-quantizing error ratio for a sinewave.

If the signal is random, there is a probability that the signal level will exceed the full scale range of the ADC. The mean square overload error is

$$\sigma_{ol}^2 = \int_{\frac{L}{2}}^{\infty} v^2 p(v)dv + \int_{\frac{-L}{2}}^{-\infty} v^2 p(v)dv \qquad (2.13)$$

where

$p(v)$ = the signal probability density
L   = the full scale ADC range

Knowledge of the statistical properties of the signal are required to compute the overload error for an arbitrary random signal. To illustrate the analysis, a Gaussian random signal is assumed with a zero mean and a mean square value, σ_s^2. The mean square overload error is

$$\sigma_{ol}^2 = 2\int_{\frac{L}{2}}^{\infty} \frac{1}{\sqrt{2\pi}\sigma_s} e^{\frac{-v^2}{2\sigma_s^2}} dv \qquad (2.14)$$

The total mean square error is the sum of the overload error and the quantization error

$$\sigma_{total}^2 = \sigma_{ol}^2 + \frac{\Delta V^2}{12} \qquad (2.15)$$

The SQR is then

$$SQR_g = \frac{\sigma_s^2}{\sigma_{ol}^2 + \frac{\Delta V^2}{12}} \qquad (2.16)$$

The SQR for the Gaussian signal is evaluated for a 10 bit ADC and plotted in Figure 2.11 as a function of the ratio of the signal RMS level to the full scale ADC range. The same general behavior is observed for the Gaussian signal as for the sinewave but the effect of overloading is much more severe.

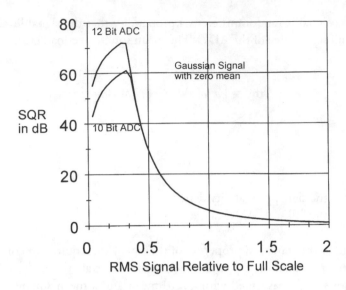

Figure 2.11 Signal-to-quantizing error ratio for a Gaussian signal.

In both cases there is an optimum ADC "loading" which maximizes perform-ance. In the case of the sinewave, the peak-to-peak amplitude should span the full scale range of the ADC. In the random signal case, the RMS signal level must be much less than full scale to minimize the overload error. The Gaussian signal assumption is a pessimistic case (a Gaussian signal is the most random signal for a given mean square value), and the performance of real signals might be expected to fall between the sinewave performance and the Gaussian signal performance.

One important feature of the SQR performance needs elaboration. Smaller amplitude signals have proportionally lower SQR. This is important in PCM telecommunication applications where user's voice levels can have a large range of levels. To place all users on a common basis, nonlinear quantization is employed with smaller quantization increments at low levels and larger incre-ments at high levels. This has been implemented using nonlinear networks known as companders as shown in Figure 2.12. Two companding functions are used in the telecommunications industry, the μ-law and the A-law. The μ-law is defined by

$$Y(x) = \mathrm{sgn}(x)\frac{\ln(1 + \mu|x|)}{\ln(1 + \mu)} \qquad (2.17)$$

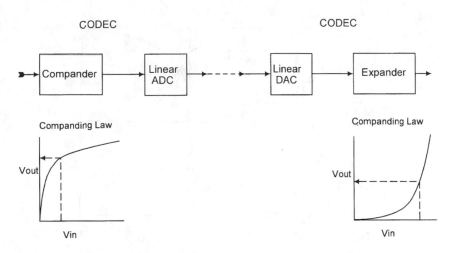

Figure 2.12 Signal companders.

while the A-law is defined by

$$Y(x) = \text{sgn}(x)\frac{1+\log(A|x|)}{1+\log A}, |x| > \frac{1}{A}$$

$$Y(x) = \text{sgn}(x)\frac{A|x|}{1+\log A}, 0 < |x| < \frac{1}{A} \tag{2.18}$$

The Bell system in North America adopted the μ-law companding using μ = 256 while much of the rest of the world uses the A-law (CCITT standard) with A = 87.5. The two companding functions are virtually identical and are shown in Figure 2.13 over a normalized range of −1 to +1. The A-law companding has somewhat smaller quantization steps at very low signal levels. Both companding laws produce nearly constant signal-to-quantization error ratios over a wide range of signal levels.

2.3 Noise and Interference

The goal of the PCM system is to transmit a series of discrete symbols from one point to another without error. The ability to do this depends upon being able to distinguish one symbol from another. In the real world, forces conspire to

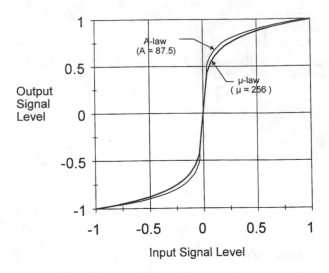

Figure 2.13 A-law and μ-law companders.

hinder the symbol decision process. These forces are noise and interference. The separation of these forces into two groups is purely arbitrary but convenient for discussion. Noise is considered to be an interfering signal which can only be described by its statistical properties. These classes of signals are also called "stochastic processes" [4]. Everything else is called "interference" and are generally quasi-deterministic signals such as unwanted clock components, crosstalk, intermodulation products, and so forth.

2.3.1 Noise

Noise is characterized by two functions, the probability density and the correlation function. For a time-varying noise function, a sample taken at a particular time has an amplitude variation described by a probability density function. Probability density functions are characterized by moments defined by

First order moment (mean value)

$$m = \int_{-\infty}^{+\infty} x \, p(x) \, dx \qquad (2.19)$$

Second order moment (mean square)

$$\sigma^2 = \int_{-\infty}^{+\infty} x^2 p(x)\, dx \qquad (2.20)$$

*N*th order moment

$$m^N = \int_{-\infty}^{+\infty} x^N p(x)\, dx \qquad (2.21)$$

The "expected value" function is defined by

$$E\{x\} = \int_{-\infty}^{+\infty} x\, p(x)\, dx \qquad (2.22)$$

As a shorthand notation, the expected value function will be used to simplify the notation. In this notation, the *N*th order moment is

$$m^N = E\{x^N\} \qquad (2.23)$$

It will be assumed in this book the noise probability density function is time-invariant, that is, does not vary from sample to sample. One of the most common assumptions about noise is that the probability density is Gaussian

$$p(x) = \frac{1}{\sqrt{2\pi}\,\sigma} \int_{-\infty}^{+\infty} e^{-\frac{(x-m)^2}{2\sigma^2}}\, dx \qquad (2.24)$$

where

σ = the root mean square value
m = the mean value

The Gaussian density is completely characterized by the first two moments, the mean and the mean square values.

The probability density tells only half of the noise story. The other half is described by the autocorrelation

$$R(t_1,t_2) = E\{x(t_1), x(t_2)\} \tag{2.25}$$

The autocorrelation is a measure of the similarity of two samples at times t_1 and t_2. If the autocorrelation depends only on the difference in time, $t_1 - t_2$, and not on the absolute values, the noise is said to be stationary.

$$R(\tau) = E\{x(t+\tau), x(t)\} \tag{2.26}$$

Although there are a few cases in PCM systems design where nonstationary noise must be considered [5], we will normally assume the noise is stationary. Some simple properties of the autocorrelation function for stationary processes are

Symmetry

$$R(\tau) = R(-\tau) \tag{2.27}$$

Maximum value

$$|R(\tau)| \leq R(0) \tag{2.28}$$

By definition, the autocorrelation at zero, $R(0)$, is the second moment of the process. If the noise is bandlimited, the autocorrelation is nonzero between samples. There is, in fact, a Fourier transform relationship between the autocorrelation function and the noise power spectrum

$$S(\omega) = \int_{-\infty}^{+\infty} R(\tau)\, e^{-j\omega\tau}\, d\tau \tag{2.29}$$

Noise which is uncorrelated except for $\tau = 0$ has a constant power spectrum. The power spectrum measures the noise power in small frequency bands as a function of frequency. A constant power spectrum is commonly called "white" noise. The power spectrum is defined over the two-sided frequency range from $-\infty$ to $+\infty$ and white noise is characterized by the spectral density, N. Most engineers tend to think only of positive frequencies and typically use the one-sided noise spectral density, $N_0 = N/2$.

When white noise is filtered, it is convenient to define an equivalent noise bandwidth such that

$$N_0 B_N = \int_0^\infty N_0 \frac{|H(j\omega)|^2}{H^2(0)} d\omega$$

or

$$B_N = \int_0^\infty \frac{|H(j\omega)|^2}{H^2(0)} d\omega \tag{2.30}$$

The equivalent noise bandwidth can be computed for many common filters. A simple RC filter has a noise equivalent bandwidth of $\pi/2$ times the -3dB frequency.

With this rather tedious background, we can now concentrate on modeling noise in PCM systems. By far the most common assumption is that system noise is white, Gaussian meaning a constant power spectrum with a Gaussian probability density. Samples of white Gaussian noise are uncorrelated and, consequently, statistically independent. This is a crucial assumption in the analysis of most PCM systems. Fortunately for most systems, the assumption of white Gaussian noise is at least a good first order approximation to reality. When noise is filtered, the samples may still be uncorrelated at certain sample periods or the noise may be "whitened" to facilitate analysis.

2.3.1.1 Thermal Noise

Noise is all around us emanating from all warm bodies. Even the lowly resistor is an important source of noise in PCM systems, maybe, the principal source. In any electrical conductor at a temperature above absolute zero the electrons are

in a state of random motion. This gives rise to thermal noise which produces an
open circuit noise voltage of

$$v^2 = 4kTR \int_{f_1}^{f_2} \frac{hf}{kT} \frac{1}{\left(e^{\frac{hf}{kT}} - 1 \right)} \, df \qquad (2.31)$$

where

> k = Boltzmann's constant = 1.38×10^{-23} joules/Kelvin
> h = Planck's constant = 6.62×10^{-34} joule-seconds
> T = resistor temperature in Kelvin
> R = resistance in ohms
> f = frequency in Hz

The complicated function under the integral is the "black body" radiation spectrum. For most systems of interest, the upper frequency, f, is such that function
is approximately unity and the integral simply evaluates to the system bandwidth, B.

$$v^2 = 4kTRB \qquad (2.32)$$

The noisy resistor can be replaced by an equivalent circuit consisting of an
ideal noiseless resistor in series with a voltage source as shown in Figure 2.14.
The available power will be defined as the maximum noise power is transferred
to a load resistor. The maximum power transfer occurs when the load resistor is
"matched" to the source resistance. For this case, it is easy to show that the
available power is kTB. By the same token, the one-sided noise spectral density
is kT. The available noise power can be used to define an equivalent noise temperature which can extend the noise power concept to sources other than a simple resistor. If the available noise from some source is equal to the noise power
from a resistor at a temperature of T_s, the noise source is said to have an equivalent noise temperature of T_s. This concept is important for very low noise systems such as communication links to deep space probes. An antenna pointed at
the night sky at C-band would see a background noise temperature of about $4K$.
The same antenna pointed at a major city might see a background noise equivalent to several hundred K.

The noise of a system can be represented by the equivalent system noise
temperature or by a figure of merit known as the noise figure. The noise figure

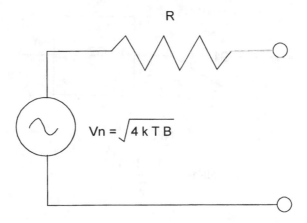

Figure 2.14 Equivalent circuit of noisy resistor.

of a network is defined as the ratio of available signal-to-noise ratio at the input of the network to the signal-to-noise ratio at the output of the network for a standard temperature of 290K. The ratio is normally expressed in decibels

$$F_{dB} = 10\log\left(\frac{SNR_{in}}{SNR_{out}}\right) \qquad (2.33)$$

The noise figure for a noiseless network is 0 dB and the noise figure is always greater than 0 dB for practical systems. The noise figure can vary with frequency of operation so that an average noise figure is often used for systems design.

If a network has an amplifier with gain, G_1 (dB), and noise figure, F_1 (dB), followed by a network with a noise figure, F_2, as shown in Figure 2.15, the effective noise figure of the combination can be computed to be

$$F_{12} = 10\log\left(f_1 + \frac{f_2 - 1}{g_1}\right) dB \qquad (2.34)$$

The effective noise figure of a cascade of networks can be extended to

$$F_{1n} = 10\log\left(f_1 + \frac{f_2 - 1}{g_1} + \frac{f_3 - 1}{g_1 g_2} + \cdots + \frac{f_n - 1}{g_1 g_2 \cdots g_n}\right) dB \qquad (2.35)$$

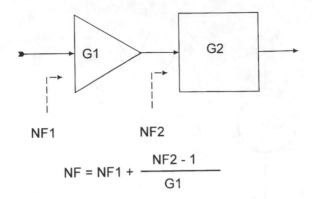

$$NF = NF1 + \frac{NF2 - 1}{G1}$$

Figure 2.15 Cascaded network.

For very low noise systems, it is more appropriate to consider the effective noise temperature of the system. Consider the simple amplifier shown in Figure 2.16 in which the amplifier has an effective noise temperature, T_{eq}. The signal-to-noise ratio at the input is

$$SNR_{in} = \frac{s_0}{kT_0 B} \tag{2.36}$$

The output signal-to-noise ratio is

$$SNR_{out} = \frac{g_1 s_0}{g_1 kT_0 B + g_1 kT_e B} = \frac{s_1}{kB(T_0 + T_e)} \tag{2.37}$$

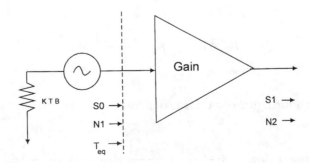

Figure 2.16 Effective noise temperature.

By definition, the noise figure is the ratio of the input signal-to-noise ratio to the output signal-to-noise ratio.

$$f = \frac{T_0 + T_e}{T_0} = 1 + \frac{T_e}{290} \qquad (2.38)$$

or

$$T_e = (f - 1)\,290 \qquad (2.39)$$

A 3 dB noise figure corresponds to an effective noise temperature of 290K. Effective noise temperature is widely used in very low noise ground stations. Parametric amplifiers have typical noise figures near 30K and lasers have been reported with noise temperatures of 10K or less. These levels approach the cosmic background levels.

In a very low noise system, the effective noise temperature of each major component in the receiver front end must be considered. The antenna receives background noise from objects, not only within the main antenna beam, but also from the antenna side lobes. At the antenna terminals this background noise has an equivalent noise temperature which is included in the overall system noise temperature. The antenna effective noise temperature is dependent on frequency, antenna pattern and on the pointing angle.

Between the antenna terminals and the first gain stage there is some loss due to either cabling and/or the insertion loss of a filter. In addition, a practical system has a certain amount of impedance mismatch. Remember, the available power assumes perfect impedance matching. A lossless mismatch reflects both signal energy and antenna noise back out of the antenna and does not degrade signal-to-noise ratio. The input termination of the low noise amplifier (LNA) propagates noise toward the antenna which is reflected back into the system by the mismatch and constitutes a noise degradation. The effective temperature degradation due to the mismatch can be computed to be

$$\Delta T = \frac{|\Gamma|^2 (T_{LNA} + T_L)}{1 - |\Gamma|^2} \qquad (2.40)$$

where

Γ = the reflection coefficient
T_{LNA} = the effective temperature of the LNA
T_L = the termination temperature

The temperature degradation can be significant for low noise systems if the mismatch is not carefully controlled. The reflection coefficient is related to the voltage standing wave ratio (VSWR) by

$$|\Gamma| = \frac{VSWR - 1}{VSWR + 1} \tag{2.41}$$

For a VSWR of 2:1 the reflection coefficient magnitude is 1/3 and the temperature degradation for a 50K LNA with a 300K termination is 43.7K, almost as much as the LNA! Reducing the VSWR to 1.5 reduces the degradation to 14.5K, a considerable improvement.

If the input has loss as well as a mismatch, the loss contributes $(L_1 - 1)\, T_{L1}$ to the system noise temperature (T_{L1} is the temperature of the lossy component). When all of these factors are considered, the effective system noise temperature is

$$T_s = T_{ant} + (L_1 - 1)T_{L1} + \frac{|\Gamma|^2 (T_{LNA} + T_L)}{1 - |\Gamma|^2} + L_1 T_{LNA} + \frac{L_1 T_{ss}}{g_1} \tag{2.42}$$

where

T_{ant} = the antenna temperature
$T_L,\ \text{T}_{L1},\ \text{T}_{LNA},\ L_1$ and Γ as defined previously
T_{ss} = the second stage noise temperature
g_1 = the LNA gain

The noise calculation in the communications worksheet discussed in Chapter 1 can be expanded to include these noise factors for a more complete analysis.

2.3.1.2 Shot Noise

Although thermal noise dominates many systems, PCM systems using optical transmitters and receivers are frequently limited by shot noise. Photo detectors exhibit a noise component known as shot noise. The current produced by light falling on the detector has a random fluctuation with a mean square value proportional to the mean current

$$i_n^2 = 2qIB \qquad (2.43)$$

where

q = the electronic charge, 1.6×10^{-19} coulombs
I = the mean photo current in amperes
B = the noise bandwidth in Hz

Shot noise limited systems exhibit an entirely different behavior than thermal noise limited systems. In the shot noise limited system, the mean square noise is proportional to the signal level so that signal-to-noise ratio is proportional to the square root of the signal level. Shot noise is associated with discrete events, namely, the emission of electrons in a photo detector in response to a stream of photons of light. The density function for shot noise is Poisson

$$p(k) = \frac{m^k}{k!} e^{-m} \qquad (2.44)$$

where

$p(k)$ is the probability of k events occurring in time, T
$m = rT =$ the average number of events in T
r = the number of events per second

The mean of this density function is m as is the mean square value. Associating an event with the arrival of a photon, if a pulse of light has an energy equivalent to 100 photons, the RMS variation is 10 photons and the signal-to-noise ratio is 10:1. Obviously in a PCM system using optical transmission, the energy in a received pulse must be much greater than one photon for a reliable system. When the mean number of photons is large, the noise is often approximated by a Gaussian density but, in general, shot noise is white Poisson noise.

2.3.2 Interference

In real systems, random noise is not always the performance limiting factor. Other interfering signals include clock signals, power supply noise, crosstalk, intermodulation distortion, and so forth. Good design practice should minimize, or eliminate these signals but Murphy's law frequently prevails and the system designer must consider their effects. Serious interference by quasi-deterministic signals can often be canceled by estimating the interfering signal and subtracting it from the received signal. This approach is often used to cancel crosstalk. When cancellation is used, the residual error is generally due to the estimation error and can be considered to an equivalent white Gaussian noise component. In fact, it can be shown that one property of an ideal estimator is that the residual error is white Gaussian.

Intermodulation distortion (IMD) can be a potential problem in PCM systems with multiplexed carriers. When two carriers are input to a nonlinear device, the output contains the fundamental components together with intermodulation products. The third order products are the strongest, can fall near the fundamental components, and cannot be easily removed by filtering. For two frequencies at f_1 and f_2, the third order products of most concern are at $2f_1 - f_2$ and $2f_2 - f_1$. Multiplexed PCM systems can have a number of frequency division multiplexed channels sharing a common RF link. If the multiplex is amplified by a transponder, or repeater, nonlinearities will produce IMD components falling in the pass bands of individual channels. For example, consider a PCM multiplex with channels spaced one MHz apart starting at 8 MHz and going to 12 MHz. The third order products from the channels at 10 MHz and 11 MHz will fall in the center of the 9 and 12 MHz channels. A satellite link is a good example of this situation. The satellite transponder typically uses a traveling wave tube (TWT) amplifier which is most power efficient near saturation. At this level, IMD is a problem and the input power level must be reduced ("backed off"). There is a tradeoff between power efficiency and performance as a function of backoff.

The third order IMD is illustrated in Figure 2.17 for a typical RF amplifier. Components are frequently specified by the third order intercept point as shown. Alternatively, the third order rejection at the nominal operating point can be used to compute the third order intercept point.

$$X = \frac{R}{2} + P_0 \ dBm \tag{2.45}$$

where

    R = the rejection of third order IMD at the nominal operating point
        in dB
    X = the third order intercept in dBm
    P_0 = the output power at the nominal operating point in dBm

As the input power is increased, the third order IMD rejection decreases 20 dB for every 10 dB increase. If the tones are unmodulated, the third order IMD will

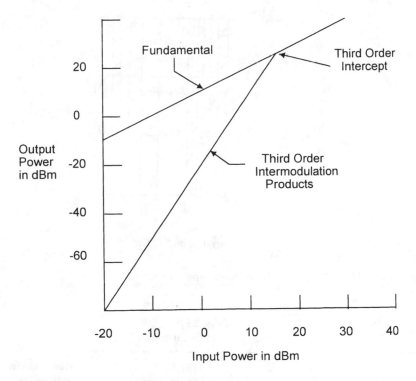

Figure 2.17 Third order intermodulation distortion.

be tones with random phase. If the tones are modulated, the IMD will approach a random interference signal.

2.4 Design Charts

The following charts summarize key system design parameters. Figure 2.18 shows the RMS aliasing error for a Butterworth (maximally flat amplitude in passband) power spectrum for orders 1 through 5 as a function of the sampling rate normalized to the −3 dB frequency. Figure 2.19 shows the maximum signal-to-quantizing noise ratio (SQR) for a sinewave as a function of the number of quantizing bits for a linear ADC. Table 2.1 summarizes the equivalent noise bandwidth for a variety of common filters. Figure 2.20 is a nomograph for rapidly computing noise power for a given noise bandwidth and noise figure.

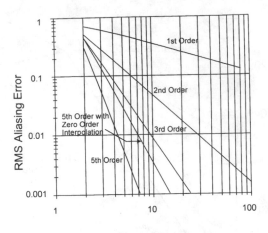

Normalized Sampling Rate

Figure 2.18 RMS aliasing error for Butterworth power spectrum.

Table 2.1
Equivalent Noise Bandwidth

Filter Type	Number of Stages	Noise Bandwidth (Relative to −3dB)
Lowpass filters		
RC	1	1.57
Butterworth	2	1.11
Butterworth	3	1.045
Butterworth	4	1.025
Butterworth	5	1.02
Bessel	2	1.1525
Bessel	3	1.073
Bessel	4	1.046
Bessel	5	1.038
Bandpass filters		
Single tuned	1	1.57
Single tuned	2	1.22
Single tuned	3	1.16
Single tuned	4	1.14
Single tuned	5	1.12
Double tuned	1	1.11
Double tuned	2	1.04

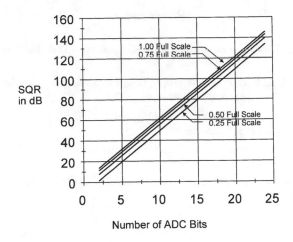

Figure 2.19 Signal-to-quantizing noise ratio for a full scale sinewave.

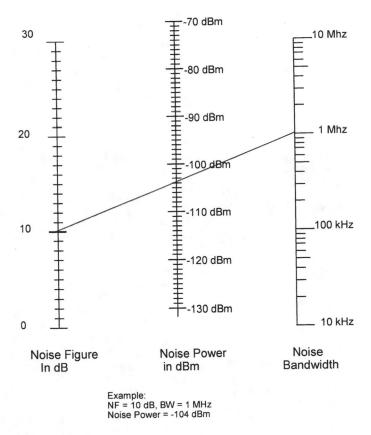

Noise Figure
In dB

Noise Power
in dBm

Noise
Bandwidth

Example:
NF = 10 dB, BW = 1 MHz
Noise Power = -104 dBm

Figure 2.20 Noise figure and noise power.

References

[1] Papoulis, A., *The Fourier Integral and Its Applications*, New York: McGraw-Hill Book Co., 1962.

[2] Gradshteyn, I. S., I. M. Ryznik, A. Jeffrey, (Editors), *Table of Integrals, Series, and Products, Fifth Edition*, New York: Academic Press, Inc., 1965.

[3] Sikora, T., "MPEG Digital Video-coding Standards," *IEEE Signal Processing Magazine*, Volume 14, Number 5, September 1997.

[4] Papoulis, A., *Probability, Random Variables, and Stochastic Processes*, New York: McGraw-Hill Book Co., 1965.

[5] Waggener, W. N., *Pulse Code Modulation Techniques*, New York: Van Nostrand Reinhold, 1995, pp 246-248.

3

Wireless Systems

In the 1990s, wireless systems have become synonymous with cellular and paging telecommunication systems. Wireless, in this book, is taken in its more general sense and includes both radio and optical systems. Although cellular systems represent a high profile example of a wireless PCM system, there are many other wireless PCM system applications. Radio telemetry and airborne data links are in widespread use. For short ranges, optical links using infrared (IR) wavelengths are also becoming more common. The television remote control using a PCM IR link is the defacto standard for controlling consumer electronic equipment.

The two types of media (radio, optical) have distinctly different characteristics as far as the PCM designer is concerned. Radio systems have the greatest diversity in characteristics. For systems using carrier frequencies greater than about 2 GHz the performance is, typically, limited by receiver thermal noise. Below 1 GHz, man-made interference and atmospheric noise often exceed receiver noise. In the band between 1 to 2 GHz, the limitation may be due to a combination of the two sources. Optical systems are frequently limited in performance by background induced shot noise in the optical detector. The distinct characteristics of each type of media is discussed in the following sections.

3.1 Radio Systems

Radio, or RF systems are probably the most common and also the oldest wireless PCM systems. The RF systems can be categorized by application as shown in Figure 3.1. The three major branches are: (1) air-to-air, (2) ground-to-air, and (3) ground-to-ground. Each of these categories can be further subdivided as shown. The RF propagation characteristics can be significantly different between the categories. A satellite-to-satellite link has nearly theoretical propagation losses while ground-to-ground link RF losses can be very hard to predict. The proximity of either the transmitter or receiver to the ground or other obstacles plays a major role in the RF link losses. Relative motion between the transmitter and receiver also plays an important role in the system performance, affecting both the link losses and the carrier Doppler shift.

This book emphasizes systems with coverage limited to an area bounded by the radio horizon as determined by the antenna heights. In a standard atmosphere, the distance to the radio horizon in miles is approximately equal to the square root of twice the antenna height in feet, a 100 ft antenna has a radio horizon of about 14 miles. In most systems, signal strength beyond the radio horizon decreases rapidly and the system performance is unacceptable. At times, VHF systems can propagate beyond the horizon due to reflections from the ionosphere. Some communication systems have been designed to use tropospheric scatter for propagation well beyond the normal horizon. These types of systems will be excluded from this book.

The basic radio link of a ground-to-air radio system is shown in Figure 3.2. The transmitter output feeds the input to an antenna which radiates an RF signal into the propagation media. The RF wave propagation is affected by the electrical characteristic of the media and by the presence of ground planes and obstacles in the path. The receiving antenna captures the radiated field and sends the received power to the receiver. The antenna also receives power from the atmosphere and other RF sources within its receiving pattern. The propagation loss can be characterized by a loss and a time-varying linear filter.

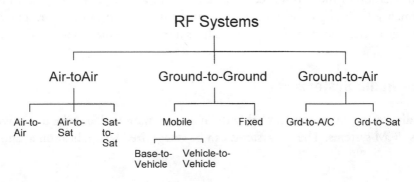

Figure 3.1 Types of radio systems.

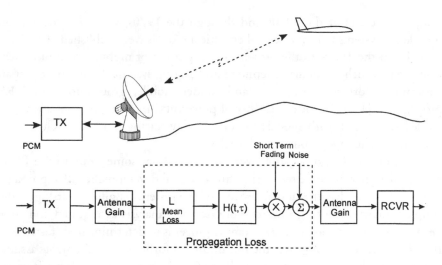

Figure 3.2 Model of RF link.

3.1.1 Radio Frequency (RF) Propagation

The development of radar and radio communications in World War II developed an intense interest in RF propagation. An enormous amount of theoretical and experimental work was conducted during this period and the classified war time results were declassified and compiled in Burrows and Atwood's book published in 1949 [1]. Bullington [2] also published an important paper in 1947 on RF propagation which is more accessible than the now out of print book by Burrows and Atwood. In the early 1950s, Reed and Russell [3] published an important book on ultra high frequency propagation which was reissued as a second edition in 1964. These works focused on basic wave propagation analysis over a spherical earth taking into account the atmosphere index of refraction, ground reflection, and diffraction by obstacles and the horizon. The theoretical work was supported by considerable experimental measurements which helped to understand the basic propagation process. Near the end of the 1950s, Egli [4] and Bullington [5] published important papers on propagation.

In the 1960s and 1970s Longley, Rice and others [6–10] pursued an ambitious project to create a large database of experimental RF propagation data and used the data to formulate a statistical model of RF propagation. The effort resulted in Fortran computer programs which allow the user to predict propagation loss for a variety of scenarios and determine a statistical estimate of the prediction accuracy. The work was so successful that the Longley-Rice model is still considered to be the de facto standard for RF propagation loss computations. At the end of the 1960s and through the 1970s, work was focused on mobile RF systems [11] and several empirical models were published [12,13].

From the 1980s to the present, the explosion of mobile telecommunication services such as cellular telephone and paging systems have resulted in an enormous effort to characterize and model mobile, ground-to-ground RF propagation [14,15]. The availability of powerful computing resources have extended single multipath models to complex ray tracing models which can include actual site topography and obstacles.

All of the RF propagation work referenced has some value to the PCM systems designer. In some sense, the chronological development of RF propagation models parallels the design process of the systems engineer starting at the simplest analysis working toward more detailed analysis as required and time permitting. The first job of the systems engineer is to determine basic feasibility. If an RF system does not meet the minimum performance requirements assuming free space loss there is "trouble in River City." Once basic feasibility is established, the RF system can be analyzed in greater detail using more complex models.

Radio systems fall into two general classes, fixed and mobile. In the fixed system, the transmitter and receiver sites are fixed in location. A microwave relay system is a good example. The system designer may be responsible for actually locating each site within a defined boundary. In this case, the designer must analyze the RF propagation for several trial sites before selecting the sites. In other cases, the site locations may be dictated by existing sites or other constraints and the designer must analyze the link propagation to predict system reliability. When the sites are fixed, ray tracing methods are suggested. With fixed links, long term weather effects are important as well as events such as sunspots. The systems engineer is often only concerned with basic feasibility and not with detailed design so a first analysis using methods discussed in later sections may be adequate, leaving a detailed link analysis to the RF engineer.

Mobile systems can be quite complex to analyze and design not only because of the motion effects (Doppler, fading), but also because of the unknown link topography. In many cases, mobile systems must operate in a mixture of environments, including urban, suburban, and rural. Furthermore, the area of operation can be flat, rolling or mountainous and tropical to subarctic. The geographical location and service area are thus extremely important. Military communication systems present particularly difficult challenges because the service area can be almost anywhere. The only way to address systems such as these is by a statistical analysis of the RF propagation. The Longley-Rice model approaches the RF propagation analysis one way while empirical models approach the analysis in a somewhat different manner.

3.1.1.1 Fundamentals

The discussion of radio wave propagation usually begins by solving Maxwell's field equations for an infinitesimal dipole in free space [1–5]. For a small dipole antenna excited by a current, I_0, as shown in Figure 3.3, the two principal field components at distance, d, are

$$E_\theta = \frac{I_0 h}{4\pi} e^{-jkd} \left(\frac{j\omega\mu}{d} + \frac{1}{j\omega\varepsilon d^3} + \frac{\eta}{d^2} \right) \sin(\theta)$$

$$H_\phi = \frac{I_0 h}{4\pi} e^{-jkd} \left(\frac{jk}{d} + \frac{1}{d^2} \right) \sin(\theta) \tag{3.1}$$

where

d = the distance
I_0 = dipole current
h = dipole length
ω = radian frequency
k = $2\pi/\lambda$
η, ϵ and μ are the magnetic and electrical properties of the media.

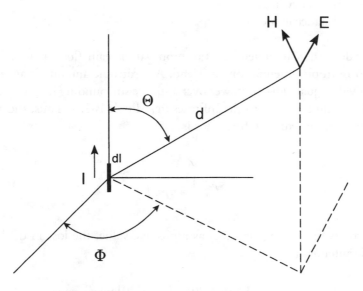

Figure 3.3 Electric and magnetic field from an infinitesimal dipole.

The wave propagation is characterized by the magnetic and electric field and the polarization. The plane of the dipole defines the plane of polarization, a vertical dipole radiates a vertically polarized wave. If two dipoles are crossed, a circularly polarized wave can be propagated. The polarization is an important parameter in RF propagation, particularly, in ground-to-ground systems.

For the region very near the dipole, the magnetic field is dominated by a term inversely proportional to the square of the distance from the dipole while the electric field term is dominated by a term inversely proportional to the cube of the distance. At longer distances from the dipole (the "far field"), both the electric and magnetic fields vary inversely with distance. Since the power is proportional to the product of the electric and magnetic field strength, the radiated power decreases with the square of the distance. The transition between the near field and the far field occurs when the distance is several wavelengths, or more. The loss between dipole radiators at one wavelength distance is about 18.6 dB.

The gain (actually a loss) between infinitesimal dipoles at a distance, d, is

$$g_{dp} = \left(\frac{3\lambda}{8\pi d}\right)^2 = \left(\frac{3c}{8\pi fd}\right)^2 \tag{3.2}$$

where

c = velocity of light = $3\ 10^8$ meters per second
f = frequency in Hz
d = distance in meters

To provide a standard reference, the propagation gain (loss) is usually referenced to isotropic antennas on each end. An isotropic antenna is an idealized antenna with equal radiated power over a sphere surrounding it. The gain of an isotropic antenna relative to the infinitesimal dipole is 2/3. Thus, the gain between isotropic antennas in free space is

$$g_{fs} = \left(\frac{c}{4\pi fd}\right)^2 \tag{3.3}$$

If we express the propagation loss as a positive value, the loss in dB between isotropic antennas in free space is

$$L_{fs} = 32.44 + 20\log(f) + 20\log(d) \tag{3.4}$$

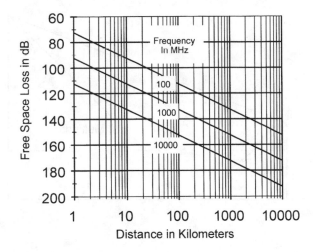

Figure 3.4 Free space propagation loss.

where

f = frequency in MHz
d = distance in km

The free space loss is shown in Figure 3.4 for frequencies of 100, 1,000, and 10,000 MHz. Notice that the loss increases at a rate of 20 dB per decade change in distance, the inverse square law.

Free space propagation can only be approached under ideal conditions. The propagation between orbiting space craft or between a ground antenna and a satellite near zenith can approach free space conditions. Thus, free space loss serves as a practical bound on performance. Under some unusual conditions, losses less than free space can be observed but these cases are a gift to the designer and cannot be relied on.

3.1.1.2 The Flat Earth Model

The free space propagation condition is one extreme of the RF link possibilities. At the other extreme, consider two antennas close to one another and to the earth. Under these conditions the earth can be considered to be flat. Rather than an electromagnetic field model, the propagation is modeled using rays. The rays are vectors normal to the wavefront with an amplitude equal to the field strength. The flat earth model is shown in Figure 3.5. Two antennas with physical heights, h_1 and h_2 are separated by a distance, d. Signals can be propagated between the two antenna via a direct ray and a reflected ray. At the point of reflection, the reflected

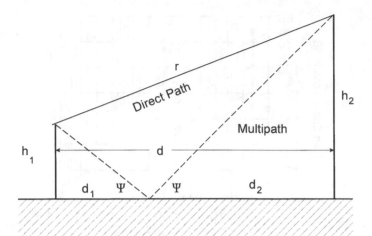

Figure 3.5 A flat earth model.

ray experiences a change of amplitude and phase which is dependent on the properties of the earth and on the polarization of the wave. The reflected wave is typically called "multipath." A surface wave can also be propagated between the two antennas which cannot be represented by the ray model but this component is negligible for frequencies above 30 MHz and is ignored. Likewise, rays can be reflected from the ionosphere but will also be ignored for this model.

Apart from the electrical characteristics, the ground reflection will also depend on the roughness of the local terrain. The ground reflected wave may be separated into a specular component and a diffuse component. The specular ray (also called "glint") obeys Snell's law with an equal angle of incidence and reflection at a single point of reflection. In addition to the specular component, some energy is scattered from the earth toward the second antenna with the largest amount originating from near the reflection point. The scattered energy is frequently called diffuse multipath.

The effect of the multipath ray on the received signal can be illustrated using phasor diagrams. Consider an unmodulated carrier represented by a phasor as shown in Figure 3.6. The specular multipath ray is represented by a phasor with an amplitude and phase shift and the resultant field is the vector sum of the direct and multipath components. The amplitude and phase of the multipath component is a function of the polarization, the frequency, the angle of incidence, and the electrical properties of the ground plane. At low angles of incidence (< 1 degree), the phase shift associated with both horizontal and vertical polarization is approximately 180 degrees, and the reflection coefficient approaches one. In addition to the phase shift due to reflection, the multipath signal has a longer path length creating a differential time delay between the

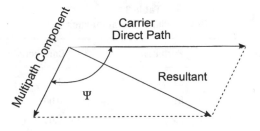

Ψ = phase lag = path difference + phase shift

Figure 3.6 Phasor diagram for an unmodulated carrier.

direct and the multipath signals. At low angles of incidence, the multipath signal can almost cancel the direct signal at some distances and reinforce the direct signal at other distances. This "lobing" effect can be quite pronounced in ground-to-air links, particularly over water.

The complex reflection coefficient plays a very important role in the propagation of signals between low antennas and near the radio horizon. The complex coefficient for horizontal polarization is

$$R_h = \frac{\sin\psi - \sqrt{n^2 - \cos^2\psi}}{\sin\psi + \sqrt{n^2 - \cos^2\psi}} = \rho_h e^{j\phi} \tag{3.5}$$

$$n = \varepsilon - j\frac{\sigma}{\omega\varepsilon_0}$$

where

ε = permitivity of the ground plane
σ = conductivity of the ground plane
ε_0 = permitivity of free space = 8.842×10^{-12}
ψ = angle of incidence
ω = radian frequency

In a similar manner, the reflection coefficient for vertical polarization is

$$R_v = \frac{n^2\sin\psi - \sqrt{n^2 - \cos^2\psi}}{n^2\sin\psi + \sqrt{n^2 - \cos^2\psi}} = \rho_v e^{j\phi} \tag{3.6}$$

Table 3.1
Ground Plane Characteristics

Type of Ground Plane	Permitivity	Conductivity (mhos per meter)
Sea water	81	1
Fresh water	81	0.001
Rich soil	20	0.03
Rocky soil	14	0.002
Sandy soil	10	0.002
Clay	13	0.004
Industrial area (city)	5	0.001

Values of permitivity and conductivity of some typical types of terrain are listed in Table 3.1.

Figure 3.7 shows the magnitude of the reflection coefficient for vertical polarization over sea water for several frequencies while Figure 3.8 shows the phase shift. The reflection for horizontal polarization has almost unity amplitude and 180° phase for angles less than 2 degrees. Figures 3.9 and 3.10 show the magnitude and phase shift for reflection of vertical polarization over a typical land ground plane. For low angles of incidence, the reflection coefficient approaches unity with a 180° phase shift for both horizontal and vertical polarization.

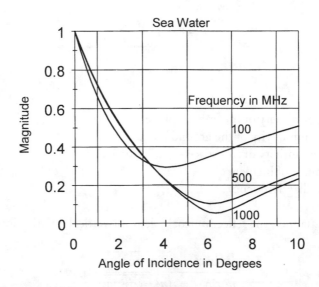

Figure 3.7 Magnitude of reflection coefficient for vertical polarization over sea water.

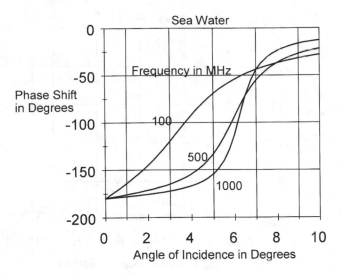

Figure 3.8 Phase shift of reflection coefficient for vertical polarization over sea water.

In addition to the phase shift due to reflection, the multipath component has an additional phase shift due to the differential delay. Using the flat earth model, the path distance for the direct ray is

$$S_{dir} = \sqrt{d^2 + (h_1 - h_2)^2} \qquad (3.7)$$

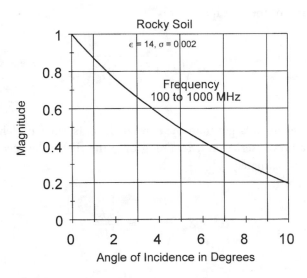

Figure 3.9 Magnitude of reflection coefficient for vertical polarization over land.

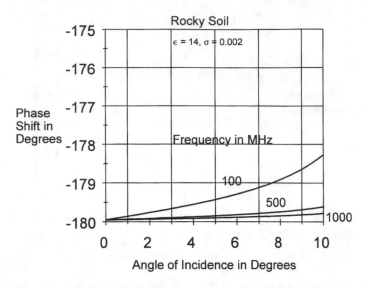

Figure 3.10 Phase shift of reflection coefficient for vertical polarization over land.

where

h_1 and h_2 are the antenna heights
d is the distance between the antennas

The path length for the multipath ray is

$$S_{mp} = \sqrt{d^2 + \left(h_1 + h_2\right)^2}$$ (3.8)

If the distance is much larger than the antenna heights, the path lengths can be expanded in a power series and, retaining only the first order terms, the path difference is approximately

$$\Delta S = S_{mp} - S_{dir} \approx \frac{2h_1 h_2}{d}$$ (3.9)

where all parameters are in consistent units.

The equivalent phase shift is

$$\Delta\theta = \frac{2\pi\Delta\theta}{\lambda} = \frac{4\pi h_1 h_2}{\lambda d} \tag{3.10}$$

The vector sum of the direct ray and the multipath ray is

$$E_r = E_d\left(1 + \rho e^{j\phi} e^{j - \frac{4\pi h_1 h_2}{\lambda d}}\right) \tag{3.11}$$

Note the sign of the reflection phase angle, ϕ, and the path difference phase shift. The reflection phase angle will be negative indicating a phase lag as is the phase shift due to the path difference. The magnitude of the received signal is

$$|E_r| = E_d\sqrt{(1 + |\rho|\cos(\phi - \Delta\theta))^2 + |\rho|^2 \sin^2(\phi - \Delta\theta)} \tag{3.12}$$

If the angle of incidence is small, the reflection coefficient is approximately -1 and the received signal is approximately

$$|E_r| \approx |E_d| 2\sin\left(\frac{2\pi h_1 h_2}{\lambda d}\right) \tag{3.13}$$

for

$$\frac{h_1 h_2}{\lambda d} \ll 1$$

$$2\sin\left(\frac{2\pi h_1 h_2}{\lambda d}\right) \approx \frac{4\pi h_1 h_2}{\lambda d} \tag{3.14}$$

The gain between isotropic antennas is approximately

$$g_{iso} = g_{fs} g_{mp} = \left(\frac{h_1 h_2}{d^2}\right)^2 \tag{3.15}$$

In dB, the loss is

$$L_{grd} = 120 - 20\log(h_1) - 20\log(h_2) + 40\log(d) \qquad (3.16)$$

where

h_1, h_2 in meters
d in km

Surprisingly, the loss for low antennas with low angles of incidence is independent of frequency. The loss with distance is 40 dB per decade as opposed to 20 dB per decade for free space propagation. Doubling the antenna heights decreases loss by 6 dB. For very low antennas ($<$ 5 meters), there is a minimum effective height for the antenna which depends on the polarization and ground plane characteristics. The net result is that the "height-gain" function for very low antenna heights is less than 6 dB per octave.

The multipath gain function, (3.13) is periodic, varying between zero and 2

$$\frac{|E_{mp}|}{|E_d|} \approx 2\sin\left(\frac{2\pi h_1 h_2}{\lambda d}\right) \qquad (3.17)$$

The function has nulls and peaks when

$$\frac{4h_1 h_2}{\lambda d} = m, m = 0,1,2\cdots \qquad (3.18)$$

With ideal specular reflection (over smooth water, for example), the propagation loss can fluctuate with distance with the loss alternating between about 6 dB less than free space to many dB greater than free space. Over more typical ground planes the "lobing" is much less pronounced and may be absent. Changing the heights of the antennas can similarly cause large changes in path loss.

The plane earth approximation is compared to data taken by the author on a UHF link over flat, sandy soil in Florida as seen in Figure 3.11. The agreement is quite good except for a short section. In this section, the line-of-sight from transmitter to receiver was partially obscured by roadside foliage. Losses

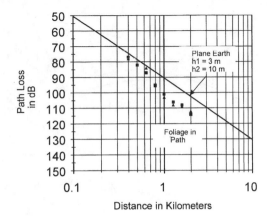

Figure 3.11 Propagation loss between low antennas.

due to foliage are discussed in a later section and the experimental losses are consistent with losses expected at the link frequency.

3.1.1.3 Spherical Earth Model

The free space and the flat earth models both have their places in the systems engineer's tool box. The spherical earth model increases the level of realism taking into account a spherical earth with an atmosphere. Ray tracing is still used for propagation within the radio line-of-sight. The effects of the atmosphere are included in the model by considering the effect of the index of refraction on the wave propagation. The velocity of propagation of an electromagnetic wave is c/n, where n is the index of refraction of the media. In the earth's atmosphere, the index of refraction varies with altitude. At high altitudes the index approaches one and the velocity of propagation approaches the speed of light. At low altitudes, the index is greater than one and the wavefront velocity is lower than the speed of light. The net result is that the wave propagation is curved by the atmosphere. Rather than dealing with curved rays, the effective radius of the earth can be modified so that the rays are straight lines in a "standard" atmosphere. For a standard earth atmosphere, the equivalent earth radius is 4/3 the actual radius. Using a 4/3 earth radius, the rays are straight lines.

The spherical earth model is shown in Figure 3.12. Two antennas with heights, h_1 and h_2, are separated by a distance, d, over a spherical earth with radius, kr, ($k = 4/3$ for standard atmosphere). There is a direct ray together with a specular multipath ray. The multipath ray strikes the surface at a distance, d_1, from h_1 and a distance, d_2, from h_2. At the reflection point, the ray amplitude

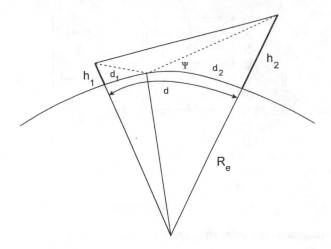

Figure 3.12　Spherical Earth model.

and phase are modified by the complex reflection coefficient as in the plane earth model. The reflection from the spherical surface causes the rays to diverge. The divergence is accounted for by a factor which multiplies the reflection coefficient. With the spherical earth model, the ratio of the received field strength to the free space field is

$$\frac{|E_r|}{|E_d|} = \sqrt{1 + DR + 2DR\cos(\theta - \phi)} \tag{3.19}$$

where

D = the divergence factor
R = magnitude of the reflection coefficient
θ = path length difference
ϕ = phase angle of the reflection coefficient

The divergence factor is computed to be

$$D \approx \left(1 + \frac{2d_1^2 d_2}{h_1' d}\right)^{-\frac{1}{2}} = \left(1 + \frac{2d_1 d_2^2}{h_2' d}\right)^{-\frac{1}{2}} \tag{3.20}$$

where

d_1, d_2 and d are the distances between antennas and the reflection point in km

h_1', h_2' are the effective heights of the antennas in meters

The effective antenna heights are

$$h_1' = h_1 - \frac{1}{2}\frac{d_1^2}{R_e}$$

$$h_2' = h_2 - \frac{1}{2}\frac{d_2^2}{R_e} \tag{3.21}$$

The maximum possible distance for direct ray transmission is called the radio horizon and for an antenna with a height of h meters is

$$d_{hor} = \sqrt{2krh} \tag{3.22}$$

for $k = 4/3$, h in meters and d in km

$$d_{hor} = \sqrt{17h} \tag{3.23}$$

The radio horizon is plotted in Figure 3.13 for $k = 4/3$, 1, and 2/3. The curves for $k = 1$ and 2/3 correspond to abnormal atmospheric conditions and illustrate how changes in conditions can change the coverage of an RF system.

To calculate the path loss for a given distance, d, requires the calculation of the ground ray intercept point. To find the point requires the solution of a cubic equation. A variety of techniques using a combination of charts and approximations have been used to find the reflection point. The author has found that a simple iterative computation which is easy and effective, using modern calculators or computers. The idea is straightforward, an approximate initial starting point is computed using the flat earth model. The angles of incidence and reflection are computed from the spherical model and the difference is used to correct the initial guess. This is iterated until the difference in the angles of incidence and reflection is less than the desired accuracy.

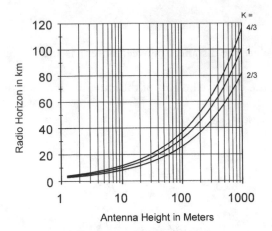

Figure 3.13 Radio horizon as a function of antenna height.

The step by step method for computing the path loss is as follows:

1) Define the antenna heights, the frequency, the polarization and the type of ground plane.

2) Assume a distance and compute the radio horizon from

$$d_{hor} = \sqrt{2krh_1} + \sqrt{2krh_2}$$

3) If the assumed distance is less than the radio horizon continue, otherwise reduce distance and go to step 2.

4) Compute the distances to the reflection point. Compute an initial estimate of d_1 from

$$\hat{d}_1 = \frac{h_1}{h_1 + h_2} d$$

5) Compute $d_2 = d - d_1$

6) Compute the effective antenna heights from

$$h'_1 = h_1 - \frac{1}{2R_e} d_1^2$$

$$h'_2 = h_2 - \frac{1}{2R_e} d_2^2$$

7) Compute the angles of incidence and reflection from

$$\psi_1 = \tan^{-1}\left(\frac{h_1'}{d_1}\right)$$

$$\psi_2 = \tan^{-1}\left(\frac{h_2'}{d_2}\right)$$

8) Compute the difference in angles. If the difference is less than the desired accuracy, proceed to step 10, otherwise, make a new estimate of d_1 using

$$d_{1new} = d_{1old} + \alpha \frac{\Delta \psi}{\psi_1} d_1$$

where

α = a constant chosen to increase (or decrease) the rate of conversion

9) Go to step six.
10) Using the angle of incidence and the ground plane characteristics, compute the complex reflection coefficient (3.5 and 3.6).
11) Compute the divergence factor (3.20).
12) Compute the phase shift due to the path difference (3.10).
13) Compute the loss relative to free space (3.19) and add (in dB) to free space loss to compute total path loss.

The method outlined in steps 1–13 are best illustrated by Example 3.1.

Example 3.1 Spherical Earth Propagation Loss

Assume a datalink between the ground and an unmanned air vehicle (UAV) flying at an altitude of 3000m. The data link is assumed to operate at a frequency of 1400 MHz with vertical polarization. The ground antenna is assumed to have a height of 10m. Compute the loss at a distance of 100 km for a ground plane with a permitivity of 10 and a conductivity of 0.002 mhos per meter.

The radio horizon is 238.9 kilometers so the link is well within the reflection region. The distance to the reflection point is computed to be

$$d_1 = 0.3322 \text{ km and } d_2 = 99.67 \text{ km}$$

then

$$h_1' \approx 10 \text{ km and } h_2' \approx 2999.4 \text{ km}$$

the reflection angles are

$$\Psi_1 = 1.5375 \text{ degrees and } \Psi_2 = 1.5375 \text{ degrees and } \Delta\Psi \approx 0,$$
$$\text{no iterations are required.}$$

The reflection coefficient is computed to be

$$\rho = 0.8358\angle -179.988 \text{ degrees}$$

The divergence factor is 0.989 so the magnitude of the reflection is 0.826. The phase shift due to the path distance is computed to be about 1009 degrees with a resultant phase shift (modulo 360 degrees) of 109 degrees. Adding the reflected ray to the direct ray as vectors result in a net vector of $1.065\angle 46$ degrees, or a power increase over free space of about 0.55 dB. In this example the multipath ray reinforces the direct ray.

3.1.1.4 Longley-Rice Model

The spherical and plane earth models provide a good starting point for the analysis of some RF links. Real links are much more complex and require more refined models. Communication links can be separated into fixed site systems, random site systems, and mobile systems. Fixed site systems contain two, or more antenna sites whose locations are known. Microwave links and some types of data acquisition systems are examples of fixed site systems. Random site systems are systems in which the locations of the antenna sites are not known a priori. A randomly sited system may have fixed antennas but the systems designer may not know where they will be located or the system may be designed to be installed in many different locales. The mobile systems typically have one, or more, fixed antenna sites communicating to mobile platforms, including cars, trucks, aircraft, satellites, and so forth. Ground-to-ground mobile systems are similar to randomly sited systems with the additional motion induced effects.

In contrast to the spherical earth model, real systems have irregular terrain profiles with vegetation, structures, and other obstacles. The atmospheric index of refraction varies with time, and altitude and weather affect propagation loss. The Longley-Rice model [6,7] and variations [8–10] of the model expand the basic spherical earth model by adding irregular terrain and other effects using semiempirical and statistical data. The Longley-Rice model is implemented in a Fortran program making it widely available. The Longley-Rice model is applicable over a frequency range from 20 MHz to 20 GHz for antenna heights between 0.5m and 3 km. Johnson and Gierhart [8–10] extended the model to include antenna heights as high as 100 kilometers provided one antenna is below 3 km and the higher antenna satisfies certain elevation conditions.

The propagation loss for both models is of the form

$$L(d) = L_{fs}(d) + A(d) + \bar{V} \tag{3.24}$$

where

$L(d)$ = the median loss as a function of distance in dB
$L_{fs}(d)$ = the free space loss as a function of distance in dB
$A(d)$ = the median excess path loss in dB
V = a random loss component with zero mean and standard deviation, σ in dB

The median excess path loss is determined semiempirically from data on a wide variety of path geometries. In the Longley-Rice model, the path loss is assumed to be lognormally distributed with a random loss component, V, with a standard deviation of σ dB. The Johnson-Gierhart model uses a two-section lognormal distribution with a breakpoint at the median value.

For most systems, three possible propagation modes may be considered:

- Free space with multipath;
- Multipath with diffraction;
- Diffraction.

The first mode is dominant when the path clears the radio horizon by greater than 1/4 of a Fresnel zone. A Fresnel zone is equal to a path difference of an integer number of half wavelengths. The Fresnel zones are illustrated in Figure

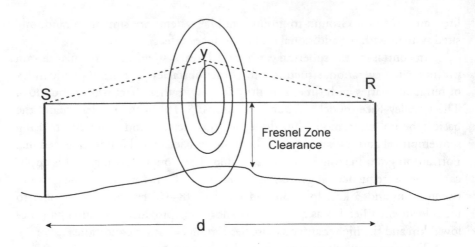

Figure 3.14 Fresnel zone clearance.

3.14. From geometry, the clearance height for a particular Fresnel zone at the midpoint of the path is

$$y = \frac{1}{2}\sqrt{n\lambda d} \qquad (3.25)$$

where

    n = Fresnel zone number
    λ = the wavelength in meters
    d = the distance between source and receiver in meters
    y = the nth zone clearance in meters

for

    d in km, f in MHz and y in meters

$$y = 273\sqrt{\frac{nd}{f}} \qquad (3.26)$$

When the path is line-of-sight but with less than then 1/4 of a Fresnel zone clearance, diffraction losses become important. The third mode (diffraction) occurs when the propagation is beyond the radio horizon by more than 1/4 of a Fresnel zone.

With random siting over irregular terrain, the line-of-sight can be obstructed even when the distance is within the nominal radio horizon. Here, diffraction becomes the dominant propagation mode. In the multipath/diffraction transition region, significant losses occur because the multipath reflection coefficient is approximately -1 at low grazing angles for all polarizations even over rough surfaces.

Three path geometries can be considered as shown in Figure 3.15. In the first case, the transmitter and receivers have direct line-of-sight. In the second case, the two antennas have a common horizon but no direct path. In the third case, two horizons are between the antennas. In the second and third cases, propagation between the two antennas is via diffraction or atmospheric scattering. In the second case, the path geometry can be modeled by a single knife-edge diffraction while the third case is modeled by two knife-edge obstacles. When the location of both sites is known, Longley-Rice computes an effective earth radius based on the altitude above sea level and the climatic conditions as expressed by the index of refraction in N-units. The refractivity at an altitude of h kilometers is

$$N_h = N_0 e^{-0.1057h} \tag{3.27}$$

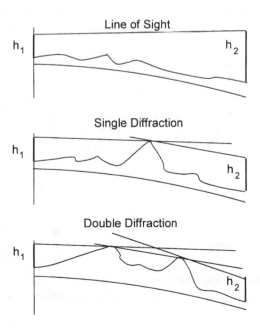

Figure 3.15 Path geometries.

where

N_0 = the refractivity at sea level in N-units

The refractivity varies with time of year, location, and local weather conditions. Minimum monthly surface refractivity referred to sea level ranges from 290 to 390, depending on location. The effective earth radius is

$$R_e = \frac{6370}{1 - 0.04665 e^{0.005577 N_b}} \qquad (3.28)$$

In the absence of specific data, the refractivity, N_b, is commonly assumed to be 301, giving an effective earth radius of 8497 km, or 4/3 actual radius.

Using the effective earth radius and the path profile, the height of the terrain above sea level is computed from the path profile corrected for the effective earth radius. A straight line is fitted to the profile and the deviation, Δh_i, of the height from the straight line is computed. The distribution of Δh_i is determined and the interdecile (0.1 to 0.9) range is defined as a measure of the terrain irregularly. Longley-Rice estimates Δh for most types of terrain as shown in Table 3.2. When an actual terrain profile is not available, or random siting of antennas is assumed, Table 3.2 can be used as an estimate of the roughness factor. The roughness factor is used to estimate the radio horizon over irregular terrain.

The median value of the horizon distance is estimated to be

$$d_{hor} = d_{1h} + d_{2h}$$

$$d_{1h} = d_1 e^{-0.07 \sqrt{\frac{\Delta h_1}{h_1}}}$$

$$d_{2h} = d_2 e^{-0.07 \sqrt{\frac{\Delta h_2}{h_2}}} \qquad (3.29)$$

where

d_1 and d_2 are the smooth earth horizons

In the Longley-Rice model, the loss, $A(d)$, is estimated at three distances, d_0, d_1 and d_2 and a smooth curve is fitted to a loss function (in dB) of the form

$$A(d) = A_0 + md + k \log_{10} d \qquad (3.30)$$

Table 3.2
Terrain Roughness Factor

Type of Terrain	Δh in meters
Water	0–5
Smooth plains	5–20
Slightly rolling	20–40
Rolling	40–80
Hills	80–150
Mountains	150–300
Rugged mountains	300–700
Extremely rugged mountains	>700

The loss is added to free space loss to determine overall loss. The three distances are chosen such that d_0 is the farthest distance that the attenuation is approximately free space. The second distance, d_1, is chosen to be well within the line-of-sight range where the two ray model is valid and the third distance, d_2, is at the spherical earth horizon. The loss at the first two distances is computed using the two ray approximation with the reflection coefficient modified by a factor

$$\rho_{LR} = \rho_{se}e^{-\left(\frac{2\pi\sigma_h \sin\psi}{\lambda}\right)} \tag{3.31}$$

where

ρ_{LR} = the Longley-Rice modified reflectivity
ρ_{se} = the smooth earth reflectivity

$$\sigma_h \approx 0.78\Delta h e^{-0.05\sqrt{\Delta h}} \tag{3.32}$$

and

$$\psi \approx \tan^{-1}\left[\frac{(h_1 + h_2)}{1000d}\right] \tag{3.33}$$

The loss at the third distance is due to diffraction and is computed as a weighted average between a smooth earth diffraction and a highly irregular earth diffraction. The weighting factor is an empirically derived function.

Some computed values of the model constants are summarized in Table 3.3 as given in [7].

Example 3.2 Longley-Rice Model

Using the computed values for the Ohio, U.S. locations with antenna heights of 4m and 9m, respectively, estimate the total propagation loss at 100 MHz at distances of 5 and 10 km. Compare the Longley-Rice estimate with the plane earth loss.

At $d = 5$ kilometers, the free space loss is

$$L_{fs}(5) = 32.44 + 20\log 100 + 20\log 5 = 86.42 \; dB$$

The excess loss is

$$L_{LR}(5) = 22.67 + 0.34787 \times 5 + 7.23686\log 5 = 10.06 \; dB$$

The total loss at 5 kilometers is 98.5 dB. The plane earth loss estimate is 116.8 dB.

At $d = 10$ kilometers, the free space loss is 92.4 dB, the excess loss is 32.3 dB for a total loss of 124.7 dB. The estimated plane earth loss is 128.87 dB.

Table 3.3
Excess Loss with Respect to Free Space Longley-Rice Model ([7])

Location	Parameters	A_0 in dB	m in dB/km	k
Colorado Plains (U.S.)	$\Delta h = 90$ $h_1 = 4\,m$ $h_2 = 6$ $f = 100\,MHz$	24.93	0.40159	7.10267
Colorado Plains (U.S.)	$\Delta h = 90$ $h_1 = 4\,m$ $h_2 = 9\,m$ $f = 100\,MHz$	22.62	0.36162	7.20867
Ohio (U.S.)	$\Delta h = 90$ $h_1 = 4\,m$ $h_2 = 6\,m$ $f = 100\,MHz$	24.98	0.38606	7.13023
Ohio (U.S.)	$\Delta h = 90$ $h_1 = 4\,m$ $h_2 = 9\,m$ $f = 100\,MHz$	22.67	0.34787	7.23686

The reader is directed to the original reports by Longley-Rice [7] and Johnson-Gierhart [8] for a complete description of the method and a listing of the Fortran programs. The limitation of the Longley-Rice model to antenna heights less than 3000m is removed by the Johnson-Gierhart model. In both cases, the reports are sufficiently complete to permit the systems engineer to customize the calculations for specific applications or to rewrite the algorithms in a different computer language.

3.1.1.5 Empirical Models

Egli Model

Many empirical models have been suggested. One model to modify theoretical calculations with empirical data was published by Egli [4]. Egli's model is based on empirical data collected by the Federal Communications Commission (FCC) obtained during studies of VHF and UHF television allocations and mobile service applications. Consequently, the model is limited to ground-to-ground systems with low (less than several hundred meters) antennas. With this limitation in mind, Egli proposed a simple model in which a frequency dependent loss term is appended to the standard plane earth formula. The empirical data suggested the loss with frequency increase at 20 dB per decade with a reference loss of 0 dB (relative to plane earth) at 40 MHz. Egli called this additional loss "terrain loss," expressed in dB as

$$L_{ter} = 20\log f - 20\log 40 \qquad (3.34)$$

When combined with the plane earth loss, the Egli model loss is

$$L_{Egli} = 88 - 20\log h_1 - 20\log h_2 + 40\log d + 20\log f \qquad (3.35)$$

Egli also observed that the empirical data was approximated by a lognormal distribution and indicated that the observed standard deviation at 127.5 MHz was about 8.5 dB, and 11.6 dB at 510 MHz. The model uses actual (physical) antenna heights with the theoretical height-gain of 6 dB per octave for heights greater than about 10m and a 3 dB per octave height-gain for heights less than 10m. There was no significant difference in the empirical data between polarizations. Based on the sites at which the data was taken, the terrain roughness factor as used by Longley-Rice is about $\Delta h \approx 150$ meters.

EPM-73 Model

Lustgarden and Madison [12] published a more complex empirical model in 1977 called the EPM-73 model. This model is based on empirical data and covers a frequency range from about 20 MHz to 10 GHz. The model is segmented into two models based on the parameter, h/λ, with one model covering "high-h/λ" and the second "low-h/λ" conditions. The high-h/λ range is considered by the authors to be greater than 25 and the low h/λ range is considered only for frequencies less than 1000 MHz and maximum antenna heights of less than 300m. When the antenna heights are unequal, the parameter, $h_1 h_2/\lambda^2$, is used to distinguish between the two models with the high-h/λ model used for $h_1 h_2/\lambda^2$ greater than about 400.

The EPM-73 model segments the range into several regions applying a loss model to each range. In the high-h/λ model, the range is divided into the reflection range, the diffraction range and the tropospheric scatter range. We will only consider the first two ranges. In the reflection region the high-h/λ model is very simple, the loss is estimated as a fixed 5 dB loss added to free space loss. In the diffraction range the loss is

$$L_{EPM} = L_{fs} + 5 + \frac{50(d - d_1)}{d_2 - d_1} \tag{3.36}$$

where

L_{EPM} = the loss in dB for high-h/λ
L_{fs} = the free space loss in dB
d = the distance in km
d_1, d_2 are computed distances in km

The distance, d_1, is the distance at which diffraction effects become noticeable and d_2 is the distance at which tropospheric scattering becomes important. These distances are computed as follows. First, several "magic" numbers are computed

$$x = h_1 h_2 f$$

$$A = \frac{2.08 \times 10^8 d_{LOS}}{10^3 - 3.75 d_{LOS}}$$

$$p = 0.6 + 1.08 \times 10^{-8} x \tag{3.37}$$

where

d_{LOS} is the theoretical line-of-sight range in km

Then, d_1 is computed

If $x \leq A$, compute

$$d_1 = \frac{1.1x}{3.47 \times 10^5} \tag{3.38}$$

otherwise, compute

$$d_1 = 1.1pd_{LOS} \tag{3.39}$$

The maximum allowable value of p is 0.9 and d_1 cannot exceed 0.99 d_{LOS}.
Empirical formulas for d_2 are

For 40 MHz $\leq f \leq$ 160 MHz

$$d_2 = d_{LOS} - 48.3 \log f + 163 \tag{3.40}$$

For $f > 160$ MHz,

$$d_2 = d_{LOS} - 16.1 \log f + 91.8 \tag{3.41}$$

The low-h/λ model is applied only for frequencies less than 1000 MHz and antenna heights less than 300m. The model is more complicated and only a simplified loss model is presented here. The reader is directed to the original paper for the exact model. The loss will only be considered for ranges greater than the cosite region (antennas in close proximity) and less than the range at which the earth's curvature need be considered. Thus, this approximation covers the same range as the plane earth models considered previously. In this region two loss models are used depending on the relative height of the two antennas
If $h_1 \gg h_2$

$$L_{EPM} = 111 - 10 \log h_1' - 20 \log h_2' + 40 \log d \tag{3.42}$$

otherwise

$$L_{EPM} = 111 - 15\log h_1' - 15\log h_2' + 40\log d \qquad (3.43)$$

where

h_1' and h_2' are the "effective" antenna heights defined by

$$h_i' = \sqrt{h_0^2 + h_i^2}$$

$$i = 1,2 \qquad (3.44)$$

The minimum effective height depends on the ground plane and the frequency. For an average ground plane, the EPM-73 approximation for the minimum effective height is

$$h_0 \approx 10^{-1.33\log f + 2.74} \qquad (3.45)$$

for

$$20 \text{ MHz} \le f \le 1000 \text{ MHz}$$

Example 3.3

Using the same parameters as Example 3.2, estimate the propagation loss using the EPM-73 model and compare the results with Longley-Rice.

$$f = 100 \text{ MHz}, \; h_1 = 4\text{m}, \; h_2 = 9 \text{ m}$$

at $d = 5$ km

$$h_0 = 1.2\text{m}, \; h_1' = 4.18\text{m}, \; h_2' = 9.1\text{m}$$

$$L_{EPM} = 111 - 15\log 4.18 - 15\log 9.1 + 40\log 5$$

$$L_{EPM} = 115.27 \; dB$$

at $d = 10$ km

$$L_{EPM} = 127.27 \; dB$$

Compared with Longley-Rice, the estimates using EPM-73 are more pessimistic at both ranges but are much closer at the 10 km distance.

One interesting aspect of the EPM-73 model is that a loss can be assumed and the distance corresponding to that loss computed. This is quite useful in doing a preliminary estimate of coverage without having to compute the loss versus distance. The original paper discusses the steps required to do the inverse computation.

Urban Models

The models considered so far, largely assume propagation in suburban or rural environments. With the advent of cellular systems, propagation in urban environments is an important aspect of the system design. Mobile radio services have been available for many years and early propagation studies were devoted to the special problems of mobile services. With the advent of cellular services driven by technology and affordable costs, interest in urban and suburban propagation estimates have increased enormously. In an urban environment, none of the models are directly applicable considering the many obstacles in the typical propagation path. The antenna heights and ranges are low enough to suggest a plane earth model. The approaches to estimation of the propagation loss and system performance are described in [13–15].

The propagation loss in the cellular environment is based purely on an empirical model. The loss model is quite simple

$$L_{cell} = L_0 + 10m \log d \;\; \text{dB} \tag{3.46}$$

where

L_0 is the loss at a reference distance, d_0
m is an empirical constant
d is the distance

The reference loss, L_0, is the loss at a predetermined short distance and may be determined empirically or may be assumed to be the free space loss at the distance. The reference distance is typically 0.1 or 1.0 km depending on the location with the shorter distance assumed in dense urban areas and the larger distance used in less dense areas. The constant, m, is empirically obtained based on measurements. In the lack of data, the constant is based on data from similar locations, for example, the measurements from New York might be used in Chicago if no local measurements are available. The constant, m, ranges from 2 (free space) to about 5 with some measured values of 4.8 (New York), 3.7 (Philadelphia), and 2.7 (several German cities.) The loss is assumed to have a lognormal distribution with a standard deviation of 6 to 10 dB.

The simple loss model has been extended to very short range systems such as a wireless link within a building. Here, the reference distance can be as short as 10m. At these short distances, near field effects can be important, depending on the frequency.

3.1.1.6 Propagation Variability

The propagation loss can be considered to be a random variable characterized by a probability distribution function. The loss has a long term and a short term variability. Empirical measurements generally fit a lognormal distribution for the long term fading component. The lognormal distribution accounts for both long term fading (over hours) and site variability. By definition, the lognormal density function is a Gaussian density with the variable expressed as a logarithm, or equivalently, as a loss in dB.

$$p(x) = \frac{1}{\sqrt{2\pi}\sigma} e^{-\frac{(x-m)^2}{2\sigma^2}} \tag{3.47}$$

where

x = the loss in dB
m = the mean loss in dB
σ = the standard deviation of the loss in dB

Then the probability that the loss is greater than some value, L, is

$$P(x > L) = \frac{1}{2} erfc\left(\frac{L-m}{\sqrt{2}\,\sigma}\right) \tag{3.48}$$

This probability is plotted in Figure 3.16.

At any particular site and time, there is also a short term fading component assumed to be either Rayleigh or Rician depending on whether the received signal is largely multipath or has a strong signal component with the multipath. Over the 10% to 90% range, the Rayleigh distribution is quite close

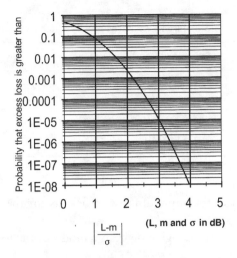

Figure 3.16 The probability that the loss is greater than *L*, dB.

to the lognormal distribution and there is a strong practical motivation to assume the short term fading is also approximately lognormal. With this assumption, the mean square short and long term variations can be added and the lognormal distribution can be used to estimate the system service quality (i.e., what is the probability that the error rate is worse than some value).

Many studies have been done of the location variability of the propagation loss for different environments. Standard deviations of loss range from 5 to 10 dB at VHF in suburban areas to greater than 15 dB in very large cities. Longley [16] summarizes many studies and suggests a model for non urban areas. Within urban areas there can be large variations depending on the characteristics of the city. Standard deviations can be very high in cities such as New York which has very high buildings while Tokyo has a much smaller deviation because of much lower building heights. The standard deviation for location variability in urban areas can be roughly modeled by

$$\sigma_L = m \log f + \sigma_0 \qquad (3.49)$$

where

    m and σ_0 are empirically obtained
    f = the frequency in MHZ
    σ_L = the standard deviation in dB

Table 3.4
Standard Deviation Model Coefficients

Environment	Frequency Band	m	σ_0	Reference
New York City	50 to 600 MHz	2.25	12	Waldo [17]
UK towns	250 to 1000 MHz	4.7	−23.6	CCIR [18]
Tokyo	100 to 3000 MHz	2.0	2	Okumura, et al. [19]
New York suburbs	127 to 510 MHz	5	−22	Egli [4]

Table 3.4 summarizes the approximate coefficients for the standard deviation for several different studies.

For nonurban areas, Longley suggests a model of the form

$$\sigma_L \approx 6 + 0.55 \sqrt{\frac{\Delta h}{\lambda}} - 0.004 \left(\frac{\Delta h}{\lambda} \right) \tag{3.50}$$

where

Δh = the terrain roughness factor in meters
λ = the wavelength in meters
σ_L = the standard deviation in dB

This approximation, shown in Figure 3.17, assumes $\Delta h/\lambda$ is less than 4700. Above this limit the standard deviation is assumed to be a constant value of 24.9 dB.

3.1.1.7 Other Losses

Several losses and other factors have been neglected in the first order analysis. First, it has been assumed that the transmitting and receiving antenna patterns have the same gain for the direct signal and the multipath signal. With highly directive antennas this is not always the case and the antenna pattern can sometimes be used to suppress multipath signals to reduce path loss. This usually is only possible with fixed sites such as microwave links and even then, weather changes can modify the multipath ray angle.

Suppression of multipath is particularly important for links with an over water path. The East and West Coast United States missile test ranges both have communication links directly over the ocean making multipath a problem at low elevation angles. Under these conditions, the links may have nearly free

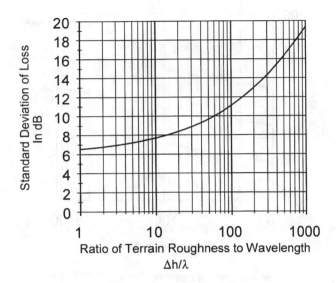

Figure 3.17 Standard deviation of loss as a function of frequency and terrain roughness [16].

space loss (20 dB per decade loss) at higher tracking angles deteriorating to 40 dB per decade or worse as the vehicles approach the horizon.

Foliage loss is another common loss neglected to this point. Trees and other foliage both absorb and diffract the RF energy. Empirical studies have lead to several loss functions. The CCIR [20] proposed a simplified foliage loss function

$$L_{foliage} = 0.2 f^{0.3} d^{0.6} \tag{3.51}$$

where

f = the frequency in MHz
d = the path distance through the foliage in meters
$L_{foliage}$ = the loss in dB

The foliage loss is plotted in Figure 3.18.

For frequencies greater than 10 GHz, atmospheric attenuation needs to be considered. Atmospheric water vapor and oxygen begin to have significant absorption at wavelengths smaller than 2 cm with a peak absorption in the 10 to 30 GHz band of about 0.2 dB per km. Very strong absorption occurs at frequencies greater than 30 GHz. In some cases, the absorption can be beneficial by limiting co-channel interference.

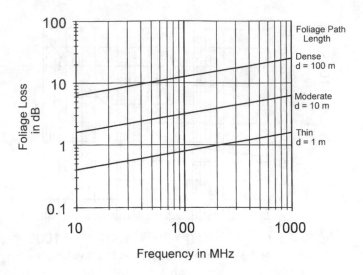

Figure 3.18 Foliage loss as a function of frequency.

3.1.1.8 Choosing the Right Model

By now, the reader should be thoroughly confused by the many different propagation models. Of the models presented, which should the systems designer choose? In clear, concise terms, the correct answer is "it depends." The choices are summarized in Table 3.5.

Although Table 3.5 specifically identifies the Longley-Rice and Johnson-Gierhart models, these represent many computer models that are available. Both commercial and shareware [21] programs are available for computing propagation losses. Some of these programs use the Longley-Rice model while others use ray tracing. The cost and availability of these programs vary.

When both antennas are low (< 100m) there are two clearly distinguished cases, urban/suburban and rural environments. The urban/suburban environment is most closely identified with cellular and personal communication systems (PCS) and the empirical urban model is suggested. Applications over water or flat land with relatively few obstructions can be reasonably approximated using a simple plane earth model. Intermediate cases are most accurately approximated using Longley-Rice although the simple Egli and EPM-73 models can provide quick estimates of loss.

When one antenna is high and one low, the EPM-73 model and a spherical earth model can provide the first order estimates with either Longley-Rice (h < 3000 meters) or Johnson-Gierhart (h > 3000 meters) providing more accu-

Table 3.5
Propagation Models

Receiving Antenna Height	Transmitting Antenna Height	
	Low	High
Low	Plane earth	Spherical earth
	Egli	EPM-73 (high h/λ)
	EPM-73 (low h/λ)	Longley-Rice (> 3000m)
	Longley-Rice	Johnson-Gierhart (> 3000 m)
	Empirical urban models	
High	Spherical earth	Free space
	EPM-73 (high h/λ)	Spherical earth
	Longley-Rice (< 3000m)	EPM-73 (high h/λ)
	Johnson-Gierhart (> 3000 m)	Johnson-Gierhart

rate estimates. The ground reflection coefficient in the spherical model can be modified to account for terrain roughness to improve the model estimate. In ground to spacecraft links, free space loss can be assumed for high ground antenna elevations while multipath loss must be considered at low elevations. Atmospheric losses need to be considered at frequencies greater than about 10 GHz. At K-band and higher, losses in rain is important.

For both high antennas, free space or the EPM-73 model (free space plus 5 dB) can be used as a first approximation. The spherical earth model can be used to approximate multipath loss. Here, the antenna directivity must be considered both in attenuating multipath as well as the direct signal. For example, in an air-to-air communication link, the directivity of the antennas can vary significantly with the aircraft attitude and link geometry.

3.1.2 Antennas

The antenna acts as the coupler between the transmitter/receiver and the propagation medium. Antennas are characterized by their gain and directivity. The propagation loss has been computed between ideal isotropic antennas that radiate equally in all directions. If we picture the transmitter with an isotropic antenna as a point source radiating equally in all directions, the power density in watts per square meter at a distance, d, is inversely proportional to the square of the distance for free space conditions. If the isotropic antenna is replaced by an antenna that concentrates the radiated energy in a beam, the received power density in that beam is increased and the antenna provides gain with respect to isotropic.

The antenna gain is defined as

$$g(\theta) = \frac{I_{max}(\theta)}{I_{average}}$$
(3.52)

where

$I_{max}(\theta)$ = the maximum radiated power in direction, θ
$I_{average}$ = the power averaged over the full hemisphere

If the antenna is assumed to have no internal losses, the gain is the equivalent to the ratio of the maximum radiated power to the radiation from an isotropic antenna. The gain is normally expressed in dB

$$G_{ant} = 10 \log g(\theta)$$
(3.53)

where

θ = the angle from the direction of maximum radiation

From reciprocity arguments, the antenna gain applies both to transmission and reception. The received power from the isotropic radiator in free space can be expressed as

$$P_{received} = \frac{J}{d^2} A_{eff}$$
(3.54)

where

J = the radiated power per unit solid angle, watts per steradian
d = the distance in meters
A_{eff} = the effective receiving antenna area in square meters

The effective receiving area can be related to the antenna gain by

$$A_{eff} = g \frac{\lambda^2}{4\pi} \qquad (3.55)$$

For the isotropic antenna, $g = 1$, and $A_{eff} = 0.08 \lambda^2$. Given the antenna gain, an effective area can be computed.

The communication systems engineer is primarily interested in the gain and directivity of the transmitting and receiving antennas. The starting point for most antenna discussions begins with the short, center fed dipole. The electric and magnetic fields resulting from a short dipole can be computed and the electric field strength in the far field varies as sin (θ), where θ is the angle from the antenna axis as shown in Figure 3.19. The maximum radiated field is at 90 degrees to the antenna. The field has a polarization in the plane of the antenna. Thus, a vertical dipole radiates vertically polarized waves and a horizontal dipole radiates horizontally polarized waves. The gain of the short dipole with respect to isotropic can be computed to be 1.5 (1.76 dB).

By extending the analysis of the short dipole, the field of a one-half wavelength dipole can be computed. The electric field in the far field is

$$E(\theta) = K \frac{\cos\left(\frac{\pi}{2} \sin(\theta)\right)}{\cos(\theta)} \qquad (3.56)$$

where

θ = the angle from the dipole axis
K = a constant

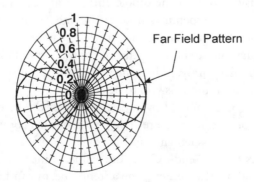

Far Field Pattern

Figure 3.19 Short dipole field.

Table 3.6
Simple Antenna Characteristics

Type of Antenna	Gain in dBi	Effective Area, λ^2	Radiation Resistance in ohms
Isotropic	0	0.08	
Short dipole	1.76	0.12	$20\pi^2\left(\dfrac{l}{\lambda}\right)^2$
$\dfrac{\lambda}{2}$ Dipole	2.15	0.13	73
$\dfrac{\lambda}{4}$ Monopole	5.16	0.26	36.5

The field is similar to the short dipole as might be expected but is somewhat more directive producing somewhat of a greater gain. The maximum radiation is perpendicular to the antenna and the antenna gain can be computed to be 1.64 (2.15 dB).

The characteristics of several simple antennas are summarized in Table 3.6. The radiation resistance is defined as that value of resistance which would dissipate the same amount of power with the same antenna current. The theoretical resistance of the half-wave dipole (73 ohms) is close to the practical value.

3.1.2.1 Antenna Arrays

Dipoles can be combined into arrays to provide greater directivity and gain. The arrays can be in a line or as a two dimensional array as shown in Figure 3.20. If the line of the array is perpendicular to the axis of the dipoles, the array is known as a broadside array. If the dipole currents are equal and in-phase, the maximum radiation is perpendicular to the plane of the array and, hence, "broadside." If the elements have equal current magnitude but a progressive phase shift, the radiation can be along the axis of the array. This type of antenna is known as an "endfire" array. When the axis of the dipole elements coincides with the array axis, the antenna is said to be collinear.

The field can be computed by superposition of the fields from each element taking into consideration the relative phases of each component. Depending on the spacing of the elements and the phasing of the currents into each element, the arrays can be "endfire" or "broadside." Since the field depends on the relative phasing of each element, the antenna beam can be steered by electronically varying the phase of each driven element. Adaptive arrays can also be

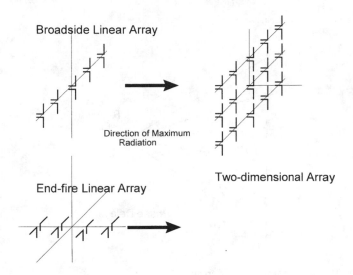

Figure 3.20 Antenna arrays.

designed which create nulls in the field pattern to cancel undesired signals. The versatility of the phased array is very important in communications applications where the directivity can be shaped and directed to optimize system perform-ance. With modern composite materials, flat panel arrays can be designed with minimal weight for weight sensitive applications.

Although the individual elements are normally dipoles, the far field from large arrays can be computed assuming the elements are isotropic radiators with little loss in accuracy. Assuming a linear array with an element spacing of d as shown in Figure 3.21, the far field is the vector sum of the intensities from each element

$$E(\theta) = \sum_{m=0}^{N-1} a_m e^{jmkd\cos\phi} \tag{3.57}$$

where

a_m = the current weighting at element, m
k = $2\pi/\lambda$

The radiation intensity is proportional to the square of the magnitude

$$P(\theta) = K|E(\theta)|^2 \tag{3.58}$$

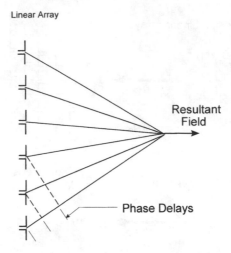

Figure 3.21 The far field from a linear array.

If all of the elements are fed with the same magnitude and phase current, the antenna pattern can be computed in closed form

$$P(\theta) = a_0^2 \, \frac{\sin^2\left(\dfrac{Nkd\,\cos(\theta)}{2}\right)}{\sin^2\left(\dfrac{kd\,\cos(\theta)}{2}\right)} \tag{3.59}$$

The antenna directivity is shown in Figure 3.22 as a function of $(kd \cos \theta)$ for an array with 16 elements. The peak response is perpendicular to the array and the main lobe width is inversely proportional to the number of elements. The width of the main lobe can be defined by the distance between the nulls on either side of the peak response. This can be computed to be

$$\Delta\theta = 2\sin^{-1}\left(\frac{\lambda}{Nd}\right) \approx \frac{2\lambda}{Nd} \tag{3.60}$$

where

    N = the number of elements
    d = the element spacing

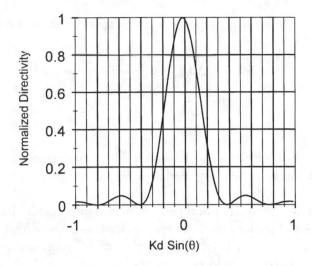

Figure 3.22 16 element array directivity.

The beamwidth is inversely proportional to the number of elements or the length of the array. The gain of the array can be computed from

$$g \approx \frac{2Nd}{\lambda} \tag{3.61}$$

The case of an array of collinear dipoles spaced one-half wavelength apart has been computed [3] and the gain is N times the gain of an individual dipole as anticipated.

3.1.2.2 Horns, Dielectric Lens, and Reflecting Antennas

Fields across an aperture can radiate energy and can act in a similar manner to dipoles. If the aperture is large, significant power can be radiated. Practical horn antennas use a waveguide to match a signal to a large aperture similar to acoustic horns. The gain of the properly designed horn is approximately

$$g_{horn} = 10\frac{A}{\lambda^2} \tag{3.62}$$

where

A = area of the horn aperture

The beamwidth is inversely proportional to the aperture size in wavelengths.

$$\Delta\theta_{horn} = C\frac{\lambda}{l}\,\text{deg} \tag{3.63}$$

where

$C = 51$ for the E-field, $l =$ the shorter side of the aperture
$C = 70$ for the H-field, $\text{l} =$ the longer side of the aperture

The equivalent of an optical lens can be designed at radio frequencies although the size of the lens normally restricts lens antennas to high frequencies. The RF lens antenna uses dielectric materials with differing indices of refraction at microwave frequencies to focus the waves.

Reflectors are much more common than lenses and use geometric optical design methods. The corner reflector and the parabolic reflector as shown in Figure 3.23 are commonly used antennas. With a half-wave dipole as a feed element, the far field is the sum of the direct wave and the reflected waves from the sides of the reflector. Gains of 10 to 12 dB are possible with the corner reflector. The parabolic reflector is often the antenna of choice when low cost and high gain is required. The parabolic reflector is limited by diffraction of the

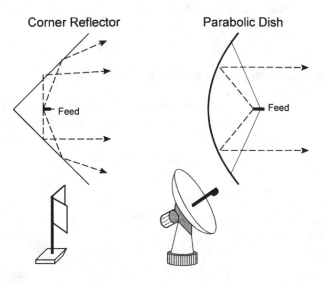

Figure 3.23 Reflector antennas.

aperture. The theoretical beamwidth between the nulls on either side of the maximum is approximately

$$\Delta\theta = 2.44\frac{\lambda}{D} \tag{3.64}$$

The theoretical gain is approximately

$$g_{par} \approx \left(\frac{\pi D}{\lambda}\right)^2 \tag{3.65}$$

In practice, an aperture efficiency factor, η, is assumed and the gain is

$$g_{par} \approx \eta\left(\frac{\pi D}{\lambda}\right)^2 \tag{3.66}$$

Assuming an efficiency of 54% (a factor based on experience), the gain expressed in dB is

$$G_{par} = 20\log D + 20\log f - 42.3 \tag{3.67}$$

where

D = aperture diameter in meters
f = the frequency in MHz

The beamwidth (between 3 dB points) is approximately [22]

$$\theta_{3db} \approx \frac{2.1\times10^4}{fD} \tag{3.68}$$

where

f = frequency in MHz
D = diameter in meters

Example 3.4 Parabolic Dish Antenna Gain

A 3m diameter antenna is required for a satellite earth station for a communication link operating in the 2.48 to 2.7 GHz band. Estimate the gain and beamwidth and compare with available antennas. The gain will be computed at 2.6 GHz. Using (3.66), the computed gain is 35.2 dB and the beamwidth is 2.7 degrees. A production antenna for this band has a specified mid-band gain of 36 dB with a beamwidth of 2.7 degrees.

For very large antenna diameters, the manufacturing tolerance limits the minimum beam size and the maximum gain.

3.1.3 Background Noise and Interference

At radio frequencies below 500 MHz, man-made and atmospheric noise can limit system performance. Figure 3.24 shows estimates of man-made noise expressed as a noise figure for frequencies from 0.1 MHz to 1 GHz based on empirical measurements [23]. Receiver noise figure limits performance above 250 MHz for quiet rural areas while man-made noise in urban environments can be significant at frequencies as high as 500 MHz. Just as the empirical propagation loss estimates are variable, the background noise level can have a large standard deviation both as a function of frequency and environment. In urban areas, the standard deviation can range from about 5 dB to about 12 dB. In residential areas, the standard deviation can range from about 5 dB to about 9 dB and in rural areas, the range is from 3 to 7 dB. Overall, the standard deviation is smallest at the high end of the spectrum (250 MHz) and largest in the 20 to 100 MHz range.

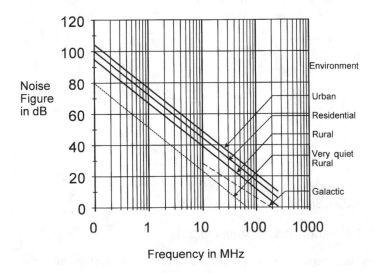

Figure 3.24 Background noise [23].

The effect of environmental noise on system performance is difficult to assess. Apart from the variability of the power level, the bandwidth and frequency distribution of the noise is frequently unknown. Man-made noise tends to have discrete frequency components with random amplitude and phase. Some types of signal modulations may be more effective than others in combating this type of noise. For systems operating in the low UHF and VHF ranges, the systems designer needs to carefully examine the environment in which the system is to operate and to allow additional design margin to accommodate the noise uncertainties.

In cellular and personal communication systems, cochannel interference is frequently the limiting performance factor. These systems are based on frequency reuse; that is, the same frequency bands are used for geographically separated channels depending on propagation loss to provide cochannel interference reduction. This concept is illustrated for a typical cellular system in Figure 3.25. Two base stations are shown for two separate regions using the same frequency channel allocations. A mobile user receives the desired signal from base station one but may also receive an interfering signal from base station two. Because of propagation loss and the greater distance to the interfering station, the interfer-

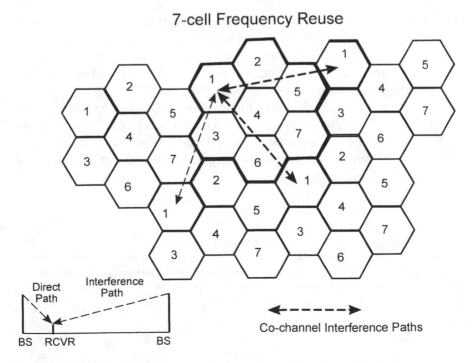

Figure 3.25 Typical cellular telecommunication system with cochannel interference.

ing signal is attenuated much more than the desired signal (that's the way it is supposed to work!). From the discussions of RF propagation loss, the empirical losses for the cellular type of system vary as

$$L(d) = K\left(\frac{d}{d_0}\right)^m \tag{3.69}$$

where

K = a constant
d_0 = a reference distance
m = a constant, typically in the range, 2 to 5

If the ratio of the distance to the interfering station to the local station is r_d, the carrier-to-cochannel interference ratio, CC/I, is

$$\frac{CC}{I} = r_d^m \tag{3.70}$$

The distance ratio depends on the frequency reuse pattern and the ratio is listed in Table 3.7 together with the CC/I for several empirical loss coefficients. This table only computes the interference due to the nearest interfering signal. In practice, the total interference from all cochannels must be included and the CC/I will be degraded from the values shown in Table 3.7.

Both the direct path loss and the interference path loss have a statistical variation and the CC/I must be described statistically. The CC/I is only meaningful when the interfering channel is on at the same time as the direct channel. There will be a certain probability that both channels are simultaneously active and cochannel interference is present.

Table 3.7
Carrier-to-Cochannel Interference

Reuse Factor	r_d	Carrier-to-Cochannel Interference Ratio in dB		
		$m = 3$	$m = 3.5$	$m = 4$
K = 4	3.46	16.2	18.9	21.6
K = 7	4.6	19.9	23.2	26.5
K = 12	6	23.3	27.2	31.1

3.2 Optical Systems

PCM using optical transmission has a long history. Ship-to-ship communication using Morse code with modulated lights is well known. Signaling using lights or smoke predates almost any other long distance communication method. The military advantage of optical communication is security. Light beams can be concentrated so that interception of the transmission is difficult. Until the introduction of the light emitting diode (LED) and the laser, optical communications were limited by the modulation capability of light sources. Modulation of light sources consisted of either mechanical shutters or modulation of the light source power. Either method limited the modulation bandwidth to very low rates.

The laser and the LED renewed interest in optical communications by providing sources with wide bandwidth and simpler modulation methods. At the same time, advances in semiconductor technology provided an entirely new set of optical detectors. Optical PCM systems have found application both in high-end military systems and in consumer remote control systems. One of the unheralded benefits of optical communications is that it spawned the new technology of "channel surfing."

In this book the optical spectrum will cover wavelengths from 0.1 to 10 μm (microns). The spectrum from 0.1 to 0.3 μm is the ultraviolet region, from 0.3 to 0.7 μm, the visible region, from 0.7 to about 2.0 μm, the near infrared (IR), from 2.0 to 5.0 μm, the middle IR region from 5.0 to 10.0 μm, and the far IR region. The atmosphere strongly absorbs bands within the optical spectrum limiting practical systems to selected windows. For systems to be used outside of the atmosphere, such as satellite-to-satellite links, optical systems are not limited by absorption considerations.

3.2.1 Propagation

Propagation losses in optical systems are due to: (1) inverse square law spreading, (2) path absorption, (3) path scattering, and (4) diffraction by obstacles. First, consider the equivalent of free space loss. Assume a point source of light (light will be used as a generic term for an optical source) which radiates equally in all directions. If the power of the light source is P_t watts, the power per unit solid angle, J, is

$$J = \frac{P_t}{\Omega} \tag{3.71}$$

where

Ω = the solid angle of the transmitted beam in steradians

At a distance, d, the power density in watts per square meter is

$$H_d = \frac{J}{d^2} \tag{3.72}$$

Using an optical receiver with an effective area of A square meters, the received power is

$$P_r = HA = \frac{P_t A}{\Omega d^2} \tag{3.73}$$

In comparison to the radio link, the reciprocal of the transmitter solid angle $(1/\Omega)$ corresponds to the transmitter antenna gain, the area, A, corresponds to the receiver antenna gain and $1/d^2$ corresponds to the free space loss. For an isotropic source, the solid angle, Ω, is 4π steradians and the received power is

$$P_r = \frac{P_t A}{4\pi d^2} \tag{3.74}$$

The basic propagation loss needs to be modified by atmospheric absorption and scattering losses along the path. The major components of the atmosphere (CO_2, CO, O_3, H_2O, and N_2O) have complex absorption bands in the optical spectrum. Although nitrogen and oxygen do not have any strong absorption bands in this region, they do affect the other absorption elements by broadening the bands. Computing the transmission losses over a real path can be quite complex. Several computer models have been developed which are widely used to compute atmospheric transmission in the optical region [24,25]. These codes generally solve Beer's equation

$$T(d) = e^{-\int_0^d \int_{\lambda_1}^{\lambda_2} k(x,\lambda)\,d\lambda\,dx} \tag{3.75}$$

where

    $k =$ the absorption coefficient per unit distance as a function of
    wavelength.

When the absorption coefficient is constant over the path and wavelength, the transmission can be simplified to

$$T = e^{-\alpha d} \tag{3.76}$$

The typical transmission over a short (16.5 km) sea level path is illustrated in Figure 3.26. This graph shows the gross structure of the transmission and is only intended to show the transmission "windows." There are usable windows in the spectral bands, 0.4 to 0.8 μm, 1.0 to 1.1 μm, 1.2 to 1.3 μm, 1.5 to 1.7 μm, 2.0 to 2.4 μm, 3.5 to 4.5 μm, and 8.0 to 12 μm. The bands of most interest for PCM communication correspond to the spectral output of LEDs and lasers. Table 3.8 summarizes the common LEDs and lasers. The red end of the visible spectrum (0.6 to 1.0 μm) is of most interest to the PCM communications designer although the 10 μm band is also of interest for long range communications because of the availability of high power CO_2 lasers.

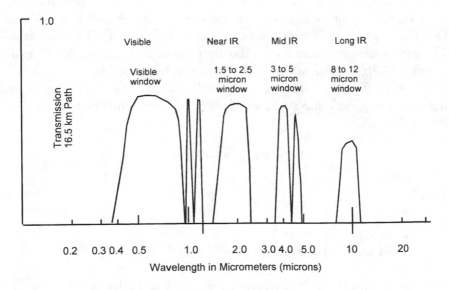

Figure 3.26 Optical spectrum transmission windows.

Table 3.8
Optical Sources

Source	Type	Wavelength (microns)	Mode	Output Power
LEDs	semiconductor	0.6 to 0.9	cw or pulsed	0.2 mw
AlGa-InP/GaAs diode lasers	semiconductor	0.63 to 0.67	cw and pulsed	0.5 to 10 mw, cw
IAl-Ga-As/GaAs	semiconductor	0.75 to 0.91	cw and pulsed	1 to 1000 mw, cw
InP/InP	semiconductor	1.06 to 1.58	cw and pulsed	1 to 15 mw
HeNe	gas	0.54 to 3.4	cw	0.1 to 50 mw, cw
HeCd	gas	0.325, 0.442	cw	5mw to 6W
CO_2	gas	9 to 11, 10.6 typical	cw	p to 15 kW
Nd:YLF	solid state	1.05, 1.3	cw and pulsed	milliwatts to watts

In addition to atmospheric absorption, scattering is an important loss mechanism. The air molecules scatter light in accordance with Rayleigh's law $(T \sim 1/\lambda^4)$ and dust, pollutants, and water vapor also cause significant scattering. The transmission with scattering is also described by Beer's law and the simplification of (3.76) is frequently used, replacing the absorption coefficient with the scattering coefficient, σ. The scattering coefficient is a function of the ratio of the scattering particle size to the wavelength. An approximation to the scattering coefficient is

$$\sigma = C_1 \lambda^{-\beta} + C_2 \lambda^{-4} \tag{3.77}$$

where

C_1 and C_2 are constants
β is an empirical constant, frequently chosen as 1.3.

The second term is the Rayleigh scattering term and is often insignificant relative to the first term. The scattering coefficient is widely variable, often ranging from 0.01 to over 1.0 km^{-1}. At the high end of the range, communication is likely to be limited to a few kilometers, or less. As a rule, it is difficult to communicate at longer distances than the eye can see. Many of the attenuation coefficients are referenced to the meteorological range. The scattering coefficient (as well as the absorption coefficient) can be expressed in terms of dB/km. At 10.6 μm, the scattering coefficient in fog (water vapor) ranges from 1 to 5 dB/km with a visibility of 1 km while the coefficient at 0.55 μm is 15 to 20 dB/km. With dust, the coefficient ranges from 5 to 15 dB/km.

Atmospheric losses are a major distinguishing difference between optical and radio systems. While multipath is important in radio systems, it is relatively unimportant in optical systems. Short range indoor systems are exceptions that use scattering and reflection to provide better coverage. Because of the susceptibility of optical systems to the vagaries of the environment, outside applications are usually very short range, specialized systems. Optical systems do have a role in space applications where propagation conditions are ideal.

3.2.2 Optical Components

Optical lenses and reflectors play the role of the RF antenna in optical systems. The optical systems generally fall into two categories: (1) refractive systems and (2) reflective systems. Refractive systems are designed using optical grade glasses (and in some cases, plastic). Reflective systems use all reflective surfaces for the optical design. There are also some hybrid (catadioptric) systems which combine reflective and refractive elements. Optical lenses are limited by the absorption characteristics of the elements and are most commonly used in the visible and near infrared regions.

In a communication system, the transmitting optical system must concentrate the transmitted power into a well defined beam with minimum power losses. The receiving system must have sufficient collecting area to provide the desired carrier-to-noise ratio. One of the advantages of an optical system is the narrow beam but this also requires precise alignment between the transmitter and the receiver.

Very few optical design concepts are required to understand the optical communication system. A simple lens, as shown in Figure 3.27, suffices to illustrate many basic principles. The key parameters of the optical system are the aperture, D, and the focal length, F. The ratio of F/D is known as the "f number" and will play an important role in the design. Abby's sine condition provides a theoretical lower bound to the f-number, $F/D \geq 0.5$. Although it is possible to approach Abby's limit, a practical bound of $F/D \geq 1$ is normally assumed.

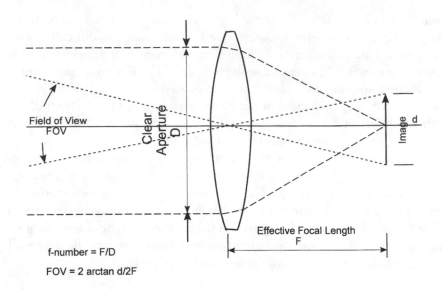

Figure 3.27 Simple optical lens.

The numerical aperture, *NA*, is an important parameter in optical communication systems. The *NA* is defined as the index of refraction of the medium containing the image multiplied by the sine of the half angle of the cone of illumination

$$NA = n\sin\theta \tag{3.78}$$

For optical systems free of spherical aberration and coma (an off-axis distortion), the *NA* is related to the *f*-number by

$$f\text{-}number = \frac{1}{2NA} \tag{3.79}$$

The power emitted by LEDs is frequently defined as the power into an optical system (a lens or optical fiber) of a given *NA*.

When a circular aperture is uniformly illuminated (as from a distant point source), the aperture diffracts the energy into a spot on the focal plane with an energy distribution of

$$I(r) = I_0 \left(\frac{2J_1(r)}{r} \right)^2 \tag{3.80}$$

where

$J_1(r)$ = the first order Bessel function
r = the radial distance in the focal plane

The image has a central peak surrounded by dark rings. The central disk is known as the Airy disk that has a diameter of

$$\Delta r = \frac{2.44 \lambda F}{D} = 2.44 \lambda \left(\frac{F}{D} \right) \tag{3.81}$$

For a practical system, the smallest image spot diameter is approximately

$$\Delta r_{min} \approx 2.44 \lambda \tag{3.82}$$

Thus, the smallest spot is about 2.5 wavelengths in diameter. The angular resolution due to diffraction is

$$\Delta \theta = 2.44 \frac{\lambda}{D} \tag{3.83}$$

The central Airy disk contains about 84% of the total image energy.

The diffraction limit is important in optical communication systems both for the receiver and the transmitter. For a system using a laser as a source, diffraction limits the smallest beamwidth achievable for a laser beam diameter. For a CO_2 laser at 10.6 μm with a beam diameter of 5 mm, the minimum beamwidth is approximately 5 mrad. As a receiver, the minimum spot size determines the minimum detector size which, in turn, determines the detector noise equivalent power.

Optical systems using lenses to focus the optical rays are known as refractive systems. Glasses have absorption bands that limit the spectral regions in which they can be used. Various fused quartz glasses can be used from about 0.3 to 3.5 μm covering the range of the semiconductor LEDs and lasers. At longer and shorter wavelengths, reflective optical systems are normally used.

Catadioptric optical systems combine a reflective mirror with a refractive "corrector" lens. A parabolic reflector provides diffraction limited performance for images on-axis but off-axis images are distorted. In addition, parabolic surfaces are difficult to manufacture. Catadioptric systems, such as the Schmidt and the Maksutov, use spherical mirrors with a refractive corrector lens to compensate for off-axis distortion. Low *f*-number systems with wide fields of view are possible with these types of systems.

For most optical communication systems, the field of view (or beam-width) can be computed assuming diffraction limiting or an angular field of view approximated by

$$\Delta\theta = 2\tan^{-1}\frac{x}{2F} \approx \frac{x}{F} \tag{3.84}$$

where

x = the detector (or source) size
F = the focal length

Example 3.5 LED Transmitter

Compute the radiated power and beamwidth from an optical transmitter using an f/1.0 lenses with a clear aperture of 10 mm. The LED specifications are

Power: 0.2 milliwatts in 0.5 NA
Emitting source size: 20 by 50 μm
Wavelength: 0.8 μm

The diffraction limit of an f/1.0 lens is

$$\Delta\theta = \frac{2.44\lambda}{D} = \frac{2.44\times0.8\times10^{-6}}{10\times10^{-3}} = 0.195\,\text{mrads}$$

The beamwidth due to the source emitting area is

$$\Delta\theta_w = \frac{w}{F} = \frac{20\times10^{-6}}{10\times10^{-3}} = 2\,\text{mrads (width)}$$

$$\Delta\Theta_h = 5\,\text{mrads (height)}$$

Thus, the beam is defined by the emitter size and not the diffraction limit and is 2 by 5 mrad or 10 microsteradians. The f/1.0 lens has a *NA* of approximately 0.5 so it is assumed that 0.2 mW is collected by the lens. If the optical lens transmittance is assumed to be 0.9, the emitted power is 0.18 mW into a solid angle of 10 microsteradians, or 18 W/sr.

3.2.3 Noise

While the receiver noise figure is generally the limiting noise in a radio system, many optical systems are background noise limited. The most common photo detectors use thermal, photo conductors, photovoltaic or photoemissive effects. Thermal detectors, such as the bolometer, are primarily used in the far infrared, depending on the energy of the signal to change the detector temperature. In a bolometer, the change in temperature changes the resistance of the detector which is sensed by a voltage divider network. Thermal detectors are generally too slow for consideration in a communication system and will not be discussed here.

The photo conductors, photovoltaic, and photoemissive detectors all directly detect the incident photons. These detectors can be quite fast and are well suited to communication applications. The physics and detailed analysis of these detectors are covered in *The Infrared and Electro-Optical Systems Handbook* [26] and only an overview is presented here.

A major difference between optical systems and radio systems is due to the quantum nature of light. Optical energy, whether from a signal or from the background, can be treated as a stream of photons. The energy of each photon is given by

$$\eta = \frac{hc}{\lambda} = \frac{1.9877 \times 10^{-19}}{\lambda} \frac{\text{watts} - \text{microns}}{\text{photons} / \text{sec}} \tag{3.85}$$

This equation answers the age old greeting, "what's new(η)?" A signal with a wavelength of 1 μm and a power of 2×10^{-19} watts corresponds to a photon flux of about 1 photon per second. When viewed as a stream of photons, the photon arrival rate is random with a Poisson distribution. The variance of the arrival rate is proportional to the average rate. If 10,000 photons are received by a detector, the standard deviation is expected to be the square root of the average, or $+/-$ 100 photons. Regardless of what sort of detector is used in an optical system, the quantum fluctuation is a fundamental limit to performance. Even with an ideal detector, the receiver must detect an average of at least one photon in a bit period. The theoretical signal-to-noise ratio is the square root of the number of photons so that the minimum energy per bit must be

$$E_b \geq \rho\eta = \frac{1.9877 \times 10^{-19} \rho^2}{\lambda} \tag{3.86}$$

where

E_b = signal energy in joules
ρ = the required signal-to-noise ratio
λ = the wavelength in microns

To achieve the theoretical bound requires an ideal photon detector with no background noise contribution or preamplifier noise. In the optical system, background energy falling on the detector also contributes to the detector noise. If the background has a steradiance of J W/sr, the background power falling in the detector is

$$P_b = \frac{\pi D^2 \tau}{4} \frac{A_d}{F^2} \int_{\lambda_1}^{\lambda_2} J(\lambda)\, d\lambda \tag{3.87}$$

where

A_d = the detector area
D = the aperture diameter
F = the focal length
τ = the optical system transmittance
$J(\lambda)$ = the background radiance in watts m^{-2} steradian

The background power is inversely proportional to the system f-number and directly proportional to the detector area. The greater the f-number, the smaller the field of view and the less background energy is admitted. An infrared detector is sensitive to any warm body near the detector and receives background energy from the surrounding optical system as well as the background within the field of view. This is important for systems operating at long wavelengths such as 10.6 μm. At wavelengths less than 1 μm, as used with LEDs and semiconductor lasers, only the background energy within the field of view is important.

Photoconductive detectors convert incident photons into free charge carriers which change the detector conductivity. The change in conductivity is converted to a voltage using a voltage divider network. The photon arrival rate is statistical so that the free charge carrier rate also follows the photon statistics. In addition to the fluctuation of conductivity due to the free charge, the free carriers also interact with the material lattice creating a generation/recombination (G-R) noise component. The detector has a "quantum efficiency" which is the ratio of the free carriers produced to the incident photons. If N_s is the number of signal photons incident on the detector and N_b is the number of background photons, the detector signal-to-noise ratio is approximately

$$\rho \approx \frac{N_s}{\sqrt{2}\left(\sqrt{N_s^2 + N_b^2}\right)} \qquad (3.88)$$

The square root of two factor accounts for the G-R noise in the photoconductive detector. This signal-to-noise ratio is a fundamental limit and preamplifier and other noise sources will further degrade the performance.

In the photovoltaic detector, photons fall on a *p-n* junction which creates a photo current proportional to the incident photon flux. The current across a *p-n* junction produces shot noise that has a noise variance

$$i_{shot}^2 = 2qI\Delta f \qquad (3.89)$$

where

q = the charge on an electron = 1.63×10^{-19} coulombs per electron
I = the total dc current across the junction
Δf = the noise bandwidth

The dc current in the detector is frequently due to the background so that the detector performance is often background limited.

The photoemissive detector emits electrons from a photocathode material that are collected by an anode. The photo current is proportional to the photon flux and has a variance equal to the shot noise. A series of multiplier stages can be added to amplify the photo current. The photo multiplier can provide current gains up to about 10^7 insuring that the detection performance is limited by the shot noise and not by external preamplifiers. The photomultiplier, although expensive, can provide near theoretical photon limited performance in the spectral region between about 0.2 and 0.9 μm. The physics of the photoemissive effect limit the upper spectral sensitivity to about 0.9 μm.

3.3 Design Charts

In many cases the system engineer needs to make quick ("back of the envelope") performance estimates of an RF link. Figures 3.28 through 3.31 provide no-

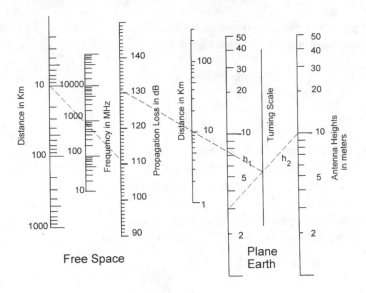

Figure 3.28 Free space and plane earth nomograph.

mographs for estimating the propagation loss for several of the simple models. Figure 3.28 computes the free space and plane earth losses. The nomograph shows one example. The example assumes a link distance of 10 km, a frequency of 1000 MHz and antenna heights of 3m and 10m. The free space loss is computed to be about 110 dB with the plane earth loss of 131 dB. The larger loss (131 dB) would be chosen as a conservative loss estimate.

The Egli approximation is computed with the nomograph in Figure 3.29. For this example the distance is assumed to be 10 km, the frequency 100 MHz, and the antenna heights, 1m and 10m. The computed loss is estimated to be 148 dB.

The EPM-73 propagation loss model for low antenna heights is computed in Figure 3.30. This example uses the same parameters as the plane earth nomograph and estimates the loss as 129.5 dB, 1.5 dB less than the plane earth model. For high antennas (or one high and one low), the EPM-73 model assumes a fixed 5 dB additional loss over free space when multipath is minimal.

In urban environments with low antenna heights (such as cellular systems), the propagation loss is computed relative to the loss at a reference distance, d_0, typically 100 m or 1 km. Figure 3.31 provides a chart to estimate the loss for several typical conditions. In the absence of a known reference loss, the free space loss at a short distance (typically, 1 km) is assumed. For example, assume a 10 km link operating at 900 MHz in an urban environment with an-

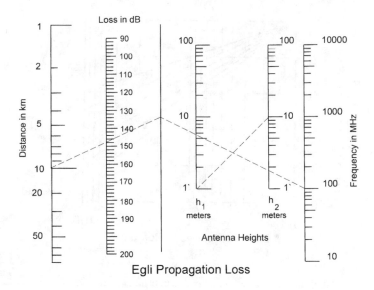

Figure 3.29 Egli propagation loss nomograph.

tenna heights of 3m and 50m. From Figure 3.28, the free space loss at 1 km is about 91 dB. Assuming the urban environment is approximated by the curve with an exponent of 3.5, the excess loss at 10 km from Figure 3.31 is about 35 dB so that the total loss is estimated to be 126 dB.

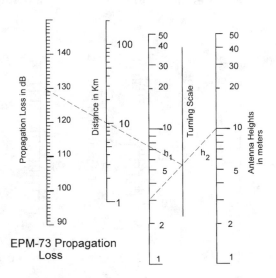

Figure 3.30 EPM-73 loss model (low antenna heights).

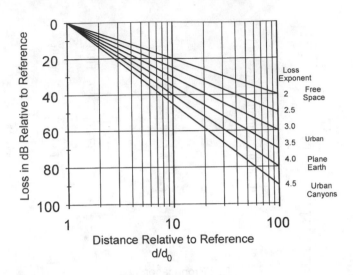

Figure 3.31 Relative loss for low antenna heights in urban environments.

In addition to the link loss, the systems engineer must also assume values for antenna gain, system losses, and receiver noise figure. The gain for some simple antennas and antenna arrays has been discussed in this chapter. The gain and beamwidth of the parabolic dish is computed in the nomograph shown in Figure 3.32. An example is shown of an antenna with a diamter of 2m operating at 1000 MHz. The gain is estimated to be 23 dB with a beamwidth of 11 de-

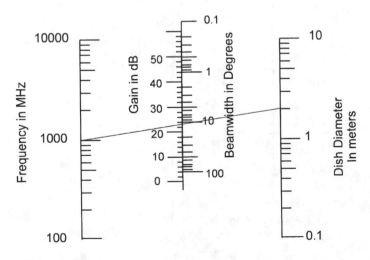

Figure 3.32 Parabolic dish antenna gain and beamwidth.

grees. The approximation computed in Figure 3.32 is reasonably accurate over the parameter range shown, although the gain is limited by manufacturing accuracy for large antennas at high frequencies.

References

[1] Burrows, C. R., and S. S. Attwood, *Radio Wave Propagation*, New York: Academic Press, 1949.

[2] Bullington, K., "Radio Propagation Above 30 Megacycles," *Proceedings of the IRE*, October 1947.

[3] Reed, H. R., and C. M. Russell, *Ultra High Frequency Propagation*, 2nd Ed., London: Chapman and Hall Ltd, 1966.

[4] Egli, J. J., "Radio Propagation Above 40 MC Over Irregular Terrain," *Proceedings of the IRE*, October 1957.

[5] Bullington, K., "Radio Propagation Fundamentals," *Bell Sys. Tech. Jour.*, Vol. 36, 1957, pp 593- 626.

[6] Rice, P. L., et al, "Transmission Loss Prediction for Tropospheric Communication Circuits," NBS Tech. Note 101, Vol. I and Vol. II, January 1967 (Available from NTIS as AD 687820 (Vol. I) and AD 687821 (Vol. II)).

[7] Longley, A. G., and P.L. Rice, "Prediction of Tropospheric Radio Transmission Loss Over Irregular Terrain: A Computer Method-1968," ESSA Tech. Report ERL-79-ITS 67 (Available from NTIS as AD 676874, 1968).

[8] Gierhart, G. D. and M. E. Johnson, "Computer Programs for Air/Ground Propagation and Interference Analysis, 0.1 to 20 GHz," Inst. Telecommunication Science (ITS) Report FAA-RD-73-103 (Available from NTIS as AD 770335), September 1973.

[9] Gierhart. G. D. and M. E. Johnson, "Application Guide, Propagation and Interference Analysis, Computer Programs (0.1 to 20 GHz)," ITS Report FAA-RD-60 (Available from NTIS as AD A053242, March 1978).

[10] Gierhart. G. D. and M. E. Johnson, "Propagation Model (0.1 to 20 GHz) Extensions for 1977 Computer Programs," ITS Report FAA-RD-77-129, (Available from NTIS as AD A055605, May 1978).

[11] Okumura, Y., E. Ohmori, and K. Fukuda, "Field Strength and its Variability in VHF and UHF Land Mobile Radio Service," *Review Electrical Communications Laboratory*, Vol 16, No. 9 and 10, 1968.

[12] Lustgarten, M. N. and J. A. Madison, "An Empirical Propagation Model (EPM-73)," *IEEE Transactions on Electromagnetic Compatibity*, Vol. EMC-19, No. 3, August 1977.

[13] Hata, M., "Empirical Formulae for Propagation Loss in Land Mobile Radio Services," *IEEE Trans. Vehic. Tech.*, Vol VT-29, No. 3, 1980, pp 317-25.

[14] Lee, William C. Y., *Mobile Cellular Telecommunications,* 2nd Edition, New York: McGraw-Hill, 1995.

[15] Sklar, B., "Rayleigh Fading Channels in Mobile Digital Communication Systems Part I: Characterization," *IEEE Comm. Mag.*, Vol. 35, No.7, July 1997, pp 90-100.

[16] Longley, A. G., "Location Variability of Transmission Loss–Land Mobile and Broadcast Systems," Report PB-254 472, NTIS, May 1976.

[17] Waldo, G. V., "Report on the Analysis of Measurements and Observations New York City UHF-TV Project," *IEEE Trans. BC*, Vol. 9, No. 2, 1963, pp 7-36.

[18] CCIR, "Influence of Irregular Terrain on Tropospheric Propagation, Filed Strength Measurements in Urban Areas for the Land Mobile Service," XIIth Plenary Assembly, Document V/136, Poland, 1969.

[19] Okumura, Y., et al, "Field Strength and Its Variability in VHF and UHF Land-Mobile Radio Service," Rev. Elect. Com. Lab., Vol. 16, 1968, pp 525-873.

[20] *CCIR XVth Plenary Assembly*, Vol. 5, Report 236-5, Geneva, 1982.

[21] *LIBRIS BRITTANIA CD ROM*, Version 4, Technical and Scientific Volume, Public Domain and Shareware Library, Crowborough, Sussex, U.K., 1994. (Available from Walnut Creek CDROM, Walnut Creek, CA).

[22] *Reference Data for Radio Engineers*, ITT Corporation, 1956.

[23] "Man-made Noise," Report 258, Int. Radio Consultative Committee, ITU, Geneva, 1980.

[24] Smith, H. J. P, "FASCODE-Fast Atmospheric Signature Code (Spectral Transmittance and Radiance,") Report AFGL-TR-78-081, Air Force Geophysics Laboratory, Jan. 1978.

[25] Kneizys, F. X., "Users Guide to LOWTRAN 7," Report AFGL-TR-88-0177, Environmental Research Papers, No. 1010, Air Force Geophysics Laboratory, Aug. 1988.

[26] Accetta, J. S., D. L. Shumacker, *The Infrared and Electro-Optical Systems Handbook*, Vol. 3, *Electro-Optical Components*, Infrared Information Analysis Center, Ann Arbor, Michigan/SPIE Optical Engineering Press, Bellingham, Washington, 1993.

4

Cable Systems

The earliest PCM systems used wire or cable as the transmission media. Cable systems continue to be important with the advent of fiber optic and coaxial wire cables. There are many applications where wireless systems are not feasible technically, economically, or politically. Except for very high frequency or very low power, most radio systems require a license from regulatory authorities to operate, and available RF spectrum is limited. In many areas the airways are becoming "polluted" with numerous wireless RF systems. Cable systems offer wide bandwidth, economic transmission links when the network sites are fixed.

Telecommunication systems probably represent the largest user of cable systems for PCM. The telephone infrastructure was largely copper wires and cable until the 1970s when fiber optic cable was introduced. The cross-over point between wire and fiber optic cable installations occurred only recently. The phone system was designed for a 3 kHz bandwidth and transmission to the home was optimized for this bandwidth. In the mid-1990s, the Internet and other digital services created an explosion of interest in providing wide bandwidth services to the home and office. Apart from the desire to provide wide bandwidth for Internet access, on-demand, pay-per-view television services were proposed requiring wideband access.

The requirement for wideband services to the user encountered what has become known as "the last mile" problem. In the early 1990s three services connected directly to the user site: telephone service, power service, and cable services. A small minority of homes had direct satellite links for television.

More recently, digital television transmission technology has greatly increased direct satellite service in competition with conventional cable systems. Of these services, cable and direct satellite provide a wideband channel to the user but do not provide a return link. The telephone system traditionally provides only voice bandwidth access and the power company provides no direct communication access. Seeing an enormous demand for higher bandwidth service, all of these service providers are trying to overcome the last mile using a variety of technologies. Although wireless technologies are addressing this problem, the majority of the systems use cable, fiber optic, or wire technology, or a combination.

Telecommunication companies began developing a broadband access standard in the 1970s as the Integrated Services Digital Network (ISDN). Despite an enormous investment in ISDN only about one million ISDN lines have been installed. The basic ISDN access provides 128 kbps which is about four times the rate achievable with voice grade modems but is much less than desired. The broadband access approach used by most telecommunication providers relies on using existing copper wires in the home with efficient cable modems. This approach is commonly known as a digital subscriber line (DSL) and has been implemented, or proposed, in several forms. Table 4.1 summarizes some of the more common DSL systems.

The telecommunication services feature switched services where each subscriber can have the full bandwidth of the access line. Cable systems have wideband access but must generally share the cable bandwidth with other subscribers in the area.

A number of system topologies have been proposed. Figures 4.1, 4.2, and 4.3 show some typical examples. These applications show the importance of a knowledge of cable technology for PCM systems engineers.

Table 4.1
Digital Subscriber Line Services

Name	Service	Rate	Mode	Remarks
IDSL	ISDN	128 kbps	Duplex	Voice & data service
HDSL	High rate DSL	1.544 Mbps	Duplex	Private T1/E1 service
ADSL	Asymmetric DSL	1.5–9 Mbps, 16–640 kbps	Downstream, upstream	Video on demand, High rate Internet access, interactive multimedia
RADSL	Rate adaptive ASDL	1 Mbps, 7 Mbps	Downstream, upstream	Same as ASDL

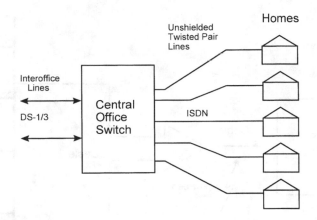

Figure 4.1 Digital subscriber loop system.

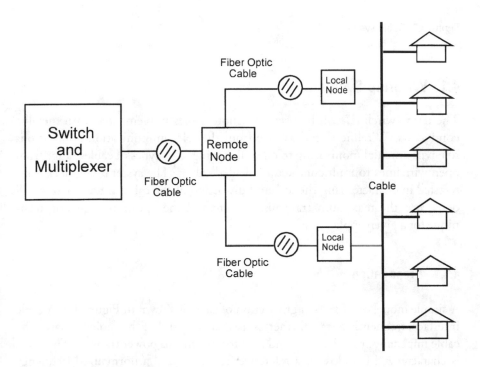

Figure 4.2 Fiber to the neighborhood system.

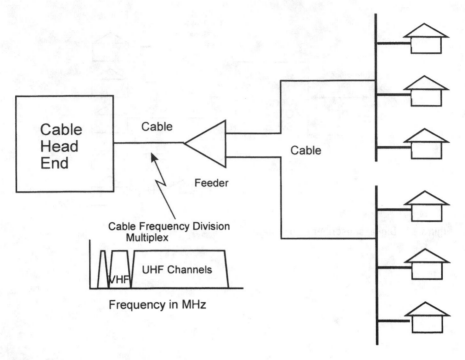

Figure 4.3 Cable system.

4.1 Wireline Systems

The term "wireline" will be used to denote cables implemented with metallic conductors. Wireline systems range from the telecommunication applications to environmental monitoring to oil well logging. The types of cables range from open wire lines to multiconductor shielded cables. The systems engineer is interested in characterizing the loss and distortion of the cable. These factors will determine the maximum transmission rate and the optimum signaling technique for a given application.

4.1.1 Propagation

A simple model of a fixed-length section of cable is shown in Figure 4.4. A cable interface, frequently a transformer, is used at each end of the cable to match the cable impedance to the source and sink for maximum power transfer. The cable is characterized by a loss at a reference frequency and a normalized frequency response which describes the amplitude and phase of the cable. As a transmis-

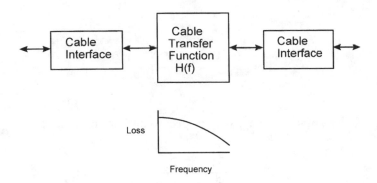

Figure 4.4 Cable model.

sion media, the cable can be highly dispersive, distorting the pulse shape of a PCM signal and introducing significant intersymbol interference.

The characteristics of a distributed transmission line can be computed theoretically for some simple cable geometries. Consider an infinitesimal short section of the transmission line as shown in Figure 4.5. The conductors have a distributed resistance and inductance along the line as well as distributed conductance and capacitance between the conductors. First, assume an ideal lossless line with R and G equal to zero. The voltage change in an infinitesimal length of line is equal to the voltage drop due to the rate of change of the current.

$$V_{x+dx} - V_x = L\,dx\,\frac{\partial I}{\partial t} \tag{4.1}$$

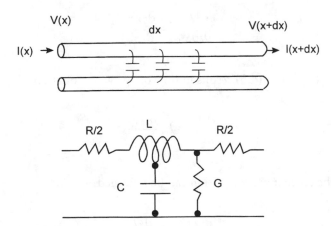

Figure 4.5 Transmission line section.

with

$$\lim_{dx \to 0} \left(\frac{V_{x+dx} - V_x}{dx} \right) = \frac{\partial V}{\partial x} \tag{4.2}$$

The partial derivative of voltage with respect to distance becomes

$$\frac{\partial V}{\partial x} = L\frac{\partial I}{\partial t} \tag{4.3}$$

In a like manner, the change in current along the line is proportional to the change in voltage

$$\frac{\partial I}{\partial x} = C\frac{\partial V}{\partial t} \tag{4.4}$$

Taking the partial derivative of (4.3) with respect to distance

$$\frac{\partial^2 V}{\partial x^2} = L\frac{\partial^2 I}{\partial t \partial x} \tag{4.5}$$

Then taking the partial derivative of (4.4) with respect to time

$$\frac{\partial^2 I}{\partial t \partial x} = C\frac{\partial^2 V}{\partial t^2} \tag{4.6}$$

Substituting (4.6) into (4.5) results in the voltage wave equation

$$\frac{\partial^2 V}{\partial x^2} = LC\frac{\partial^2 V}{\partial t^2} \tag{4.7}$$

Similarly, the current wave equation can be obtained

$$\frac{\partial^2 I}{\partial x^2} = LC\frac{\partial^2 I}{\partial t^2} \tag{4.8}$$

The velocity of propagation can be defined as

$$v_p = \frac{1}{\sqrt{LC}} \tag{4.9}$$

The characteristic impedance of the line will also be defined as

$$Z_0 = \sqrt{\frac{L}{C}} \tag{4.10}$$

A solution to the wave equation can be found to be

$$V(x) = f\left(t - \frac{x}{v_p}\right) \tag{4.11}$$

The reader can verify that this is a solution by substituting it into the wave equation. This solution represents a wave traveling along the line with velocity, v_p. It can also be shown that a second solution is a wave traveling the opposite direction so a complete solution is the sum of the two waves

$$V(x) = f\left(t - \frac{x}{v_p}\right) + f\left(t + \frac{x}{v_p}\right) \tag{4.12}$$

If a pulse is applied to the ideally lossless line it will be propagated without distortion until a line discontinuity is encountered. Suppose the line is terminated in a resistance, R. The current in the pulse will initially be V_0/Z_0. At the terminating resistor, the voltage will be V_0 and the current in the resistor would be V_0/R. Unless $R = Z_0$, the resistor current will not equal the line current and the system will compensate by launching a reflected wave such that the voltages and currents are in balance. Thus, if the line is not terminated in the characteristic impedance, pulses will be reflected at each mismatch distorting the signal and causing "echos."

Now consider a lossy line with distributed resistance and conductance. Rather than taking a general approach with arbitrary time-varying signals, the line response to a sinusoidal signal is examined. The complex frequency of the signal is represented by the complex Laplace variable, s. Following the same approach used for the lossless line, the change in voltage along the line is equal to the voltage drop across the line resistance and inductance

$$\frac{\partial V_x}{\partial x} = (R + sL)I_x \tag{4.13}$$

$$\frac{\partial I_x}{\partial x} = (G + sC)V_x \tag{4.14}$$

Combining these equations, the wave equations are

$$\frac{\partial^2 V_x}{\partial x^2} = \gamma^2 V_x \tag{4.15}$$

$$\frac{\partial^2 I_x}{\partial x^2} = \gamma^2 I_x \tag{4.16}$$

where

$$\gamma^2 = (R + sL)(G + sC) \tag{4.17}$$

is defined as the propagation constant and

$$Z_0 = \sqrt{\frac{R + sL}{G + sC}} \tag{4.18}$$

is the characteristic impedance. For the lossless line, the characteristic impedance is real (resistive) while, in general, the impedance is complex. The propagation constant is also complex.

The solution to (4.15) and (4.16) is

$$V_x = V_0 \cosh(\gamma x) - I_0 Z_0 \sinh(\gamma x) \tag{4.19}$$

$$I_x = I_0 \cosh(\gamma x) - \frac{V_0}{Z_0} \sinh(\gamma x) \tag{4.20}$$

Generally, the line conductance is essentially zero and at low frequencies the resistive loss is much less than the inductive reactance. For these conditions, the propagation constant is approximately

$$\gamma = \alpha + j\beta \approx \omega\sqrt{LC} \cdot \sqrt{-1 + j\frac{R}{\omega L}} \tag{4.21}$$

Equations (4.19) and (4.20) are in the same form as an *h*-parameter two-port network [1]. The *h*-parameters for the distributed line are

$$h_{11} = -Z_0 \sinh(\gamma z)$$

$$h_{12} = \cosh(\gamma x)$$

$$h_{21} = \cosh(\gamma x)$$

$$h_{22} = -\frac{1}{Z_0}\sinh(\gamma x) \tag{4.22}$$

Two-port analysis can be used to analyze the frequency response of a section of cable.

Example 4.1 Cable Characteristics

Compute the propagation constant, characteristic impedance and *h*-parameters for a 100 m length of coaxial cable with the following parameters at a frequency of 1 MHz.

$C = 80$ pF/m, $L = 0.2$ μH/m, $R = 0.026$ ohms/m, $G = 0$ mho/m

The propagation constant is computed from

$$\gamma = \omega\sqrt{LC}\sqrt{-1 + j\frac{R}{\omega L}}$$

$$\gamma = 7.896 \times 10^{-2}\sqrt{-1 + j0.021} \approx j0.079$$

$$Z_0 = \sqrt{\frac{R + j\omega L}{j\omega C}} = \sqrt{\frac{0.026 + j1.257}{j5.03 \times 10^{-4}}}$$

$$Z_0 = 50 \text{ ohms}$$

$$\gamma x = j0.079 \times 100 = j7.9$$

$$h_{11} = -50 j \sin 7.9 = -j49.95$$

$$h_{12} = h_{21} = \cos 7.9 = -0.046$$

$$h_{22} = -j0.02 \sin 7.9 = -j0.199$$

The transmission lines described so far are idealized, constant parameter models. In practice, the distributed parameters are frequency dependent. The line resistance increases with frequency due to the skin effect. At higher frequencies the current in the line becomes concentrated near the surface of the cable. Theoretically, the resistance increases as the square root of frequency. The skin resistance per unit length [2] is

$$R_s = \sqrt{\frac{\pi f \mu}{\sigma}} \tag{4.23}$$

The skin resistance as a function of frequency for copper and aluminum cable is shown in Figure 4.6. The cable attenuation is approximately

$$L_{cable} \approx 4.343 \left(\frac{R_s}{Z_0} + GZ_0 \right) dB / m \tag{4.24}$$

For most cables, the conductance can be assumed to be zero so that the cable loss is proportional to the square root of frequency

$$L_{cable} \approx \alpha \sqrt{f} \tag{4.25}$$

where

α = an empirical constant

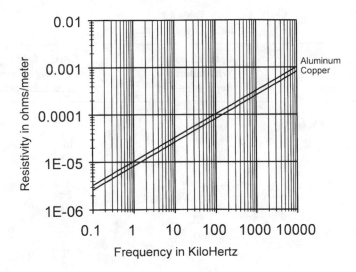

Figure 4.6 Skin resistance of copper and aluminum cables.

4.1.2 Cable Characteristics

The characteristics (R, L, C, G) of several cable configurations have been computed theoretically. Table 4.2 summarizes the theoretical parameters for an open pair wire and a coaxial cable. The characteristic impedance, Z_0, for the two cases are given by

Parallel conductors

$$Z_0 = \frac{276}{\sqrt{k}} \log_{10} \frac{d}{r} \qquad (4.26)$$

where

k = the dielectric constant of the media
d = the separation of the conductors
r = the radius of the conductors

Coaxial cable

$$Z_0 = \frac{138}{\sqrt{k}} \log_{10} \frac{r_1}{r_0} \qquad (4.27)$$

Table 4.2
Cable Characteristics

	Open Wire Cable, Wire radius = r, Separation = d	Coaxial Cable, Inner wire radius = r_0, Outer cable radius = r_1
Capacitance, C, F/meter	$\dfrac{\pi\varepsilon}{\cosh^{-1}\left(\dfrac{d}{r}\right)}$	$\dfrac{2\pi\varepsilon}{\ln\left(\dfrac{r_1}{r_0}\right)}$
Inductance, L, H/meter	$\dfrac{\mu}{\pi}\cosh^{-1}\left(\dfrac{d}{r}\right)$	$\dfrac{\mu}{2\pi}\ln\left(\dfrac{r_1}{r_0}\right)$
Conductance, G, mho/m	$\dfrac{\pi\sigma}{\cosh^{-1}\left(\dfrac{d}{r}\right)}$	$\dfrac{2\pi\sigma}{\ln\left(\dfrac{r_1}{r_0}\right)}$
Resistance, R, ohms/m	$\dfrac{2R_s}{\pi d}\dfrac{\dfrac{d}{r}}{\sqrt{\left(\dfrac{d}{r}\right)^2-1}}$	$\dfrac{R_s}{2\pi}\left(\dfrac{1}{r_1}+\dfrac{1}{r_0}\right)$
Characteristic impedance	$\dfrac{1}{\pi}\sqrt{\dfrac{\mu}{\varepsilon}}\cosh^{-1}\left(\dfrac{d}{r}\right)$	$\dfrac{1}{2\pi}\sqrt{\dfrac{\mu}{\varepsilon}}\ln\left(\dfrac{r_1}{r_0}\right)$
Attenuation in dB/meter	$4.343\left(\dfrac{R_s}{Z_0}+GZ_0\right)$	$4.343\left(\dfrac{R_s}{Z_0}+GZ_0\right)$

where

k = the dielectric constant of the media
r_1 = the separation of the conductors
r_0 = the radius of the inner conductor

The cables of interest for PCM communication are usually twisted pair lines for short distances and coaxial cables for long distances. At PCM bit rates of a few thousand bits per second or less, the RS-232D standard is defined for multiwire cables. The RS-232 standard specifies bipolar signal levels of +/−5 to +/−15 volts and is usually driven by standard digital logic drivers. The RS-232 standard can be used at distances up to about 20m, at rates up to about 30 kbps, and as long as 100m for low rates. For higher rates, the RS-422/423 standard was developed as a data transmission for short range communication

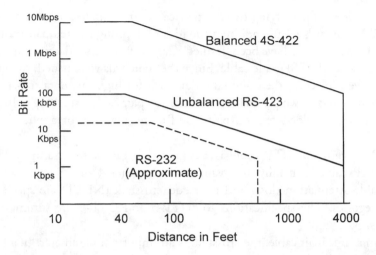

Figure 4.7 Data rate versus distance.

over twisted pair lines at rates up to 10 Mbps. The data rate versus distance envelope for the data transmission standards are shown in Figure 4.7. Rates up to 10 Mbps at short distances can be achieved using balanced RS-422 lines with commercially available drivers and receivers. A typical high speed RS-422 driver and receiver circuit is shown in Figure 4.8. The balanced line with differential drive is very effective at eliminating common mode interference.

Within the last five years, twisted pair cables have made enormous advances in data transmission and are challenging coaxial cable in many high bit rate applications. Historically, little attention was paid to the control of the impedance parameters of twisted pair cables (such as the lowly telephone pairs.) In the late 1970s, special versions of twisted pair cables were designed which had much better performance for data transmission. These early efforts eventually

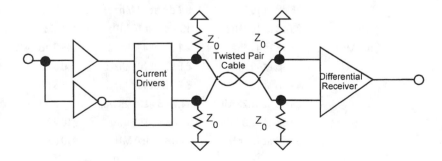

Figure 4.8 RS-422 cable driver and receiver.

led to the idea of classifying twisted pair cables into grades, or categories. The TIA/EIA standards groups adopted the idea of defining twisted pair cable categories. Five categories have been defined. Category 1 is essentially plain old telephone service (POTS) grade cable. Since the standards were issued, categories 3 and 5 have evolved as the dominant standards for high rate data transmission. Although there are two general types of twisted pair cables, shielded twisted pair (STP) and unshielded twisted pair (UTP), the majority of applications use UTP.

A category m UTP cable is designated UTP-m, for example, UTP-3. The UTP cables generally are available with either 4 or 25 pairs per cable. The cable attenuation (loss) and near end crosstalk (NEXT) are specified for the categories. The specifications for UTP-3 and UTP-5 are summarized in Table 4.3.

In a multipair cable, the crosstalk is an important parameter. In a full duplex system (simultaneous transmission in both directions), the received power on a pair of lines is the transmitted power minus the cable attenuation. Crosstalk from adjacent pairs (NEXT) interferes with the received power. For equal transmitter powers on each end, the NEXT is equal to the transmitter power minus the NEXT attenuation. Thus the ratio of NEXT attenuation to the cable loss (ACR) is a figure of merit which can either estimate the maximum cable

Table 4.3
UTP Cable Specifications

Category	Bandwidth	NEXT (minimum)	Attenuation (maximum per 100 meters)	ACR at 100 m (attenuation to crosstalk ratio) *see note*
3	16 MHz	41 dB at 1 MHz		
		32 db at 4 Mhz	5.6 dB at 4 Mhz	26.4 dB
		26 dB at 10 Mhz	9.7 dB at 10 Mhz	16.3 dB
		23 dB at 16 Mhz	13.1 dB at 16 MHz	9.9 dB
5	100 MHz	53 dB at 4 Mhz	4.1 dB at 4 Mhz	48.9 dB
		47 dB at 10 Mhz	6.5 dB at 10 Mhz	40.5 dB
		44 dB at 16 Mhz	8.2 dB at 16 Mhz	35.8 dB
		39 dB at 31.25 Mhz	11.7 dB at 31.25 Mhz	27.3 dB
		35 dB at 62.5 Mhz	17 dB at 62.5 Mhz	18 dB
		32 dB at 100 MHz	22 dB at 100 MHz	10 dB

Note: The ACR is not part of the specification.

length for a given bandwidth or the maximum bandwidth for a given cable length. In some systems, active crosstalk cancellation can minimize the NEXT component but not eliminate it entirely.

The UTP standard defines the cable characteristic impedance, Z_0. For UTP-5, the standard specifies an impedance of $100 +/-15$ ohms. With UTP, the characteristic impedance can vary significantly and some cable manufacturers must rely on the use of a measurement smoothing formula to meet the impedance specification. The UTP cables are balanced so that baluns are required to interface the cables with unbalanced coaxial sources and loads. Designing and building the balun is not a trivial task at 100 MHz bandwidth and can be a problem.

Unlike the RS-422 data transmission systems, the high rate UTP cable systems require cable modems to achieve multi-megabit per second transmission. The choice of symbol encoding and modulation for UTP cables is an area of concern for the PCM designer. The tradeoffs will be discussed later but the nature of the design problem for high rate systems over cable can be illustrated as follows. Signals can be designed to convey one bit per symbol, two bits per symbol, and so forth. As an example, a 10 Mbps signal can be transmitted with one bit per symbol requiring a bandwidth of roughly 10 MHz. If a four level symbol is used, two bits can be transmitted with each symbol and the same 10 Mbps can be transmitted using a 5 megasymbol per second rate requiring roughly 5 MHz bandwidth. It takes more noise margin to distinguish four levels than two levels but the cable loss at 5 MHz is less than at 10 MHz. At the same time, NEXT increases with frequency and will be more severe at 10 MHz. These tradeoffs suggest an optimum design for a given cable application.

Coaxial and fiber optic cables are normally used for the transmission of high rate PCM signals over long distances. Fiber optic cables are considered in Section 4.2. The wireline cable loss is proportional to the product of the square root of the bandwidth and the cable length. The bandwidth is proportional to the PCM bit rate so the maximum cable length is inversely proportional to the square root of the bit rate. The cable bandwidth is a function of the cable size, larger cables have less loss and greater bandwidth. The cables used for cable television can support a bandwidth of nearly one GHz for short spans.

Although coaxial cables are the dominant technology for many wide bandwidth applications, there are applications which use specialized, multiconductor cables. Examples of multiconductor cables, shown in Figure 4.9, include a balanced two conductor, shielded cable, often called twinaxial (twinax), and other specialized cables as illustrated. Theoretical analysis of these cables is difficult and the multiple conductors can support multiple modes of transmission. If there are N conductors and a transmission circuit is formed

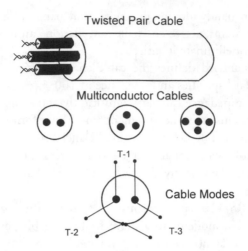

Figure 4.9 Multiconductor cables.

by two conductors, the number of possible circuits is given by the combinatorial equation

$$m = \frac{N!}{2(N-2)!} \tag{4.28}$$

If the shield on the two conductor, shielded cable is considered as a third conductor, three circuits can be formed; one is the primary balanced circuit while the other two are between each conductor and the shield. In a five conductor cable, 10 circuits are possible.

Not all of the possible transmission circuits are useful, or desirable. The transmission modes are typically separated using transformer coupling to minimize coupling and interference between circuits. The different circuits can have dramatically different propagation characteristics due to the differing inductance and capacitance of the wire pairs. The characteristics of the different circuits, in most cases, must be determined empirically. Despite the differences, most of the circuits exhibit a transfer function of the following form

$$H(f) = K_1 e^{-a\sqrt{f}}\, e^{-j\sqrt{LC}} \tag{4.29}$$

Empirical data can be used to determine the transfer function constants. Figure 4.10 shows an example of the frequency response of one circuit in a long, seven conductor cable. A least squares curve fit to the cable loss fits the theoretical

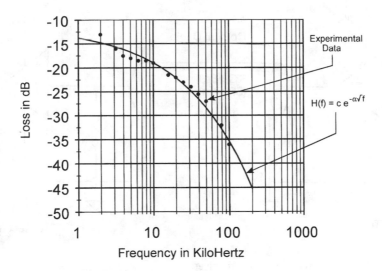

Figure 4.10 Experimental cable loss.

square root of frequency with a correlation coefficient of better than 0.99. The analytical approximation can be used to design cable equalization circuits and to estimate the time domain response.

4.1.3 Cable Drivers and Receivers

The typical cable driver and receiver for twisted pair lines was illustrated earlier. Coaxial cables can be driven in a similar manner as shown in Figure 4.11. Standard digital logic drivers and receivers can drive RG58/U coaxial cable over 1000m at rates up to about 100 kbps and at rates to 20 Mbps for shorter lines. High speed emitter coupled logic can drive coaxial cables at multi-megabit per second rates. For long lines, more sophisticated transmitters and receivers are required. Ultimately, the maximum distance for a given cable and data rate is determined by the receiver noise figure in the same manner as radio systems.

4.2 Fiber Optic Systems

Optical fiber technology started in the 1960s but began to rapidly expand in the 1970s with the production of 20 dB/km single mode fiber. Since that time, the bandwidth-distance parameter for fiber optics has increased about an order of magnitude every five years [3]. The long line telecommunication link offers a good example of the evolution of a fiber optics application [4]. In 1970, a 100

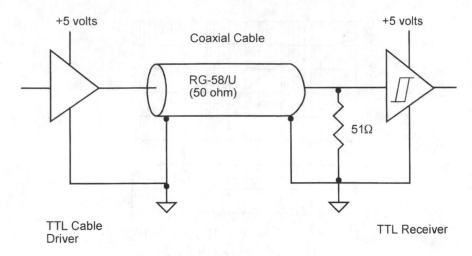

Figure 4.11 Coaxial cable drivers and receivers.

km coaxial cable link at 140 Mbps required 50 repeaters and had a mean time to failure (MTBF) of about 0.4 years. In the decade from 1970 to 1980, a 100 km, multimode optical fiber link required 10 repeaters and had an MTBF of about 2 years. By the 1980s only 3 repeaters were required and the MTBF increased to about 10 years. At the present time no repeaters are required for a 100 km link and the MTBF is about 100 years.

At the systems level, there is very little mystery about fiber optics technology. The construction of an optical fiber is shown in Figure 4.12. A very small diameter glass fiber is surrounded by a cladding material, also glass. This is the basic

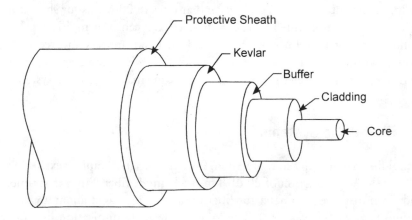

Figure 4.12 Typical optical fiber cable.

transmission media. Additional materials surround the glass fiber to provide mechanical and environmental protection. When light enters one end of the fiber it is contained within the central core by reflection from the cladding material until it emerges from the end of the fiber. From a systems standpoint, the cable parameters of interest are the end-to-end loss and the cable dispersion, or bandwidth.

4.2.1 Propagation

Propagation of light in the optical fiber can be treated by geometrical ray tracing. When a ray goes from a medium of refractive index, n_1, to a medium with index, n_2, the ray is refracted according to Snell's law as shown in Figure 4.13.

If the second medium has a lower index of refraction than the first, the ray will be refracted at a larger angle than the angle of incidence. At a critical angle, known as Brewster's angle, the ray will be totally reflected internally. If we consider a fiber optic cable, shown in Figure 4.14, there is an angle at which rays entering the fiber will intercept the cladding at the Brewster angle and will be totally reflected internally. This angle is known as the numerical aperture, NA, of the fiber. The NA can be computed from the refractive indexes of the core fiber and the cladding by

$$n_1 \sin\theta_1 = n_2 \sin\theta_2 \tag{4.30}$$

$$NA = \sin^{-1}\sqrt{n^2_{core} - n^2_{clad}} \approx \sqrt{n^2_{core} - n^2_{clad}} \tag{4.31}$$

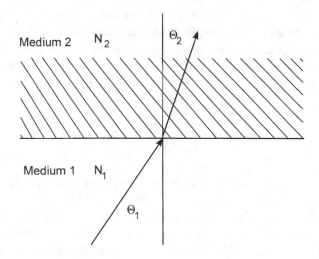

Figure 4.13 Snell's Law.

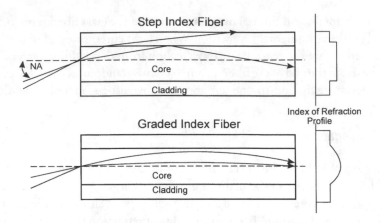

Figure 4.14 Fiber optic cable numerical aperture.

Example 4.2 Fiber Optic Cable Numerical Aperture

Compute the numerical aperture for an optical fiber with an index of refraction of 1.48 with a cladding index of 1.46.

$$NA = \sin^{-1}\sqrt{1.48^2 - 1.46^2} = 0.244$$

Once a ray enters the fiber, the primary loss is absorption by the core material. When the core diameter is large compared to the wavelength of light, off axis rays within the critical angle propagate along the fiber. From electromagnetic field theory, as the core diameter approaches a few wavelengths, propagation is by guided waves and higher order modes are attenuated. From a geometric viewpoint, only the axial ray propagates along the fiber. The number of electromagnetic wave modes which can be propagated is approximately related to the ratio of the core diameter to the wavelength and the numerical aperture by

$$N_m \approx \frac{\pi^2}{2}\left(\frac{D}{\lambda}\right)^2 NA^2 \qquad (4.32)$$

For a fiber with a $NA = 0.21$ (12 degrees), a single mode is supported for a fiber diameter less than about 2.1 wavelengths. In practice, fibers with diameters of 5 to 10 μm will support single mode propagation with lasers in the 0.85 to 1.3 μm wavelength range. Fiber optic systems are classified as single mode fiber (SMF) or multiple mode fiber (MMF) according to the allowable propagation. In general, SMFs have core diameters less than 10 μm while MMFs have core diameters in the 100 to 200 μm range. Fiber optic cable manufacturers typically describe the cables in terms of the core diameter/overall diameter in μm, for example, 100/140.

In a MMF, a transmitted pulse is broadened as it is propagated due to the different path lengths of the multiple rays. This time dispersion (also known as modal dispersion) is a function of the index of refraction profile. A SMF only propagates the axial ray so there is no modal dispersion. In addition to modal dispersion, the index of refraction varies with wavelength so that a spread in spectral energy in the source will also lead to a time dispersion. This is known as chromatic dispersion.

The simplest fiber geometry uses a core with one index and a cladding with a lower index. This geometry is called a step index fiber. The path length in the fiber is not equal for all rays and the difference in propagation time is proportional to the difference in index of refraction between the core and the cladding. The difference in time is

$$\Delta t = \frac{S}{c}\left(n_{core} - n_{clad}\right) = \frac{S\Delta n}{c} \tag{4.33}$$

where

S = the length of the fiber
c = the speed of light

For a step index fiber with $\Delta n = 0.015$, the maximum time dispersion is 50 nanoseconds per kilometer. If the time dispersion is Gaussian, the fiber bandwidth is approximately

$$B \approx \frac{0.312}{\Delta t} = \frac{0.312c}{S\Delta n} \tag{4.34}$$

The bandwidth-distance product is a constant equal to

$$B \times S = \frac{0.0936}{\Delta n} \; MHz - km \tag{4.35}$$

This product is a conservative estimate since the maximum time delay was used to estimate bandwidth.

The modal dispersion in an MMF can be dramatically improved by varying the index of the refraction profile of the core fiber. If a parabolic variation in

index from the center of the fiber to the cladding is used, the time dispersion is
reduced to

$$\Delta t \approx \frac{S}{2c} \Delta n^2 \qquad (4.36)$$

and

$$B \times S \approx \frac{0.1872}{\Delta n^2} \, MHz - km \qquad (4.37)$$

Again using a $\Delta n = 0.015$, the maximum time dispersion is 0.375 nanoseconds
per kilometer, a huge improvement in bandwidth.

The chromatic dispersion depends on the fiber material and the spectral
width of the optical source. A typical LED has a spectral width in the 50 to 100
nanometer range. The index of refraction change for fused quartz is about
$-0.018 \, \mu m^{-1}$ at 0.8 microns so the change in velocity of propagation is about
one part in 1000. This is about an order of magnitude less than the change in a
step index fiber but can be the limiting dispersion in a graded index fiber. The
chromatic dispersion can be minimized by operating near 1.3 μm where the
rate of change of index is small.

The loss in the fiber is dependent on the core material and its purity. A
quartz glass is typically used and the loss per kilometer is shown in Figure 4.15
in the 0.8 to 1.5 μm range. The limiting loss factor is scattering within the fiber.
Rayleigh scattering varies inversely with the fourth power of wavelength so operating at longer wavelengths is desirable. Other losses are due to absorption by
OH and other impurities within the glass. There are three distinct windows
which are used in fiber optic systems. These windows are defined partly by the
fiber loss and partly by available sources. The band from 0.8 to 0.9 μm is attractive because it is the region where low cost LEDs radiate. Typical losses for glass
fiber in this region are 2 to 2.5 dB per kilometer. The other regions of interest
are around 1.3 and 1.5 μm where LED sources are available and fiber losses are
low.

Plastic fibers [5] are also available for short optical links where the cost of
the fiber is important. Local area networks (LANs) are an ideal application for
plastic optical fibers (POF). Plastic optical fibers have high losses but also have
large core sizes with large numerical apertures. The large core areas reduce the
sensitivity to alignment and reduce *NA* loss. The POFs have very good LED
coupling and low cost connectors can be used. The POF is primarily based on

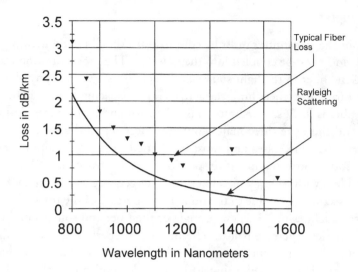

Figure 4.15 Glass fiber loss.

polymethyl methacrylate (PMMA) which has a minimum loss window at 0.58 μm. Losses in this window are typically 100 to 150 dB per kilometer although the theoretical limitation is in the order of 10 dB per kilometer. The 0.65 μm region is better matched to available LEDs although the loss is 160 to 200 dB per kilometer. Despite the high loss, the POF can be applied to short links (100 m).

A comparison of the different types of fiber optic cables is summarized in Table 4.4.

Table 4.4
Fiber Optic Cables

Type of Fiber	Core/Clad Diameter	Numerical Aperture	Loss dB/km	Bandwidth-distance MHz-km
Glass single mode (SMF)	5-10/80-125	0.1	0.2	$> 10^5$
Glass multimode (MMF)	50/125, 85/125,	0.2–0.3	3–10	20–100
Step mode glass	110/140			
Multimode (MMF)	50/125, 62.5/125,	0.2–0.3	0.5–3.0	400–1000
Graded index plastic	100/140			
Multimode	460/500, 960/1000	0.47	160	6

4.2.2 Couplers

Light from the transmitter must be coupled into the fiber and the power at the receiving end must be coupled into the detector. The most common method of coupling the fiber to the light source and detector is by direct physical contact. Manufacturing tolerances limit the actual contact spacing but it will be assumed that the fiber is in close proximity to the detector and light source. At the transmitting end there are three major sources of loss. First, the fiber core must be larger than the source area and must be aligned. If the overlapping area is less than the source area, the loss is proportional to the loss in area. This is usually not a problem with MMF which have much larger core sizes than SMF.

The second loss is due to the numerical aperture of the fiber. The numerical aperture defines the acceptance angle of the fiber and source energy outside of the numerical aperture is lost. The numerical aperture loss depends on the angular distribution of the power from the light source. There are three common sources: (1) surface emitting LED, (2) edge emitting LED, and (3) laser diode. The surface emitting LED is approximated by a Lambert power distribution

$$P(\theta) = P_0 \cos\theta \tag{4.38}$$

The edge-emitting diode emits a narrow beam in one dimension ($<$ 1 degree) with a broad (20 to 40 degrees) beam in the second dimension. The loss in the one dimension is less than that incurred with the LEDs so the numerical aperture is less.

The power distribution for LED devices can be generalized to a function

$$P(\theta) = P_0 \cos^m \theta \tag{4.39}$$

The coupling loss can be computed by integrating these distributions over the numerical aperture of the fiber. The loss in dB as a function of numerical aperture is shown in Figure 4.16 for a Lambertian emitter and a typical range for available LEDs. The larger numerical aperture of POF ($\sim$0.4) is an advantage for short links where the higher cable loss can be tolerated.

The third coupling loss is due to reflection at the fiber surface. This loss is typically less than 0.5 dB and can be ignored or considered part of the numerical aperture loss.

Because of the precision required to minimize coupling losses, manufacturers of LEDs and laser diodes provide "pigtails" which are coupled closely to the emitting area and provide an external fiber optic interface. In most cases, the

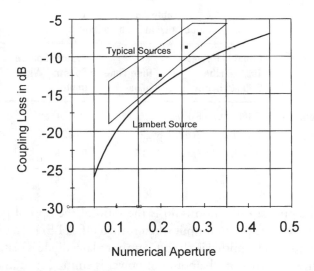

Figure 4.16 Numerical aperture loss.

manufacturer of the source provides data on the actual power coupled into a fiber of a given numerical aperture eliminating the need to estimate the loss.

At the receiving end, light is emitted from the fiber in a cone approximately equal to the numerical aperture. If the detector area is large enough to cover the divergent cone of light, the coupling loss is essentially zero. If the detector area is small, an optical lens may be used to image the light onto the detector. In this case, losses due to the lens should be included, although the losses can normally be ignored. Pigtails are also commonly used on detectors designed for fiber optic applications.

4.2.3 Transmitters

The three most common transmitters for fiber optic cable systems are: (1) surface emitting LEDs, (2) edge-emitting LEDs, and (3) injection laser diodes. The two categories of LEDs are distinguished by the emitting surface. The surface emitting devices radiate from the surface with approximately Lambert distribution. The edge-emitting diodes confine the emission to a waveguide which emits along the edge of the device. By confining the emission, the edge devices have less numerical aperture loss than the surface-emitting diodes. The injection laser diode is a semiconductor with a geometry which forms a waveguide cavity permitting stimulated emission.

The LEDs offer high reliability, low cost, and high optical extinction ratio. Laser diodes, on the other hand, are used when higher power and faster speed is

Table 4.5
Semiconductor Light Emitting Diodes

Source	Coupled Power 100/140 Fiber 0.29 *NA* in mw	Rise Time in ns	Typical Source Spectral Width in nm	MTBF (25° C)
LED surface emitting	0.160 mw, −8 dBm	2.5	80-120	10^6
LED edge emitting	0.45 mw, −3.5 dBm	2.5	80-120	10^7
Laser diode	> 1 mw, 0 dBm	0.5	0.1–1.5	10^6

required. The optical extinction ratio is the ratio of the optical power output when turned on to the power output when turned off. A LED can be completely turned off so no light is emitted in the off state but a laser diode is biased just below the lazing threshold so a small amount of power is emitted even in the off state. Table 4.5 compares the LED and injection laser diode characteristics.

4.2.4 Detectors

The two most common types of detectors for fiber optic communication systems are *p-i-n* photo diodes and avalanche photo diodes (APDs.) The *p-i-n* photo diode uses a thin *p+* material exposed to the light within an antireflection coating. Photons can penetrate the material into an intrinsic layer where electron-hole pairs are created. The diode is back biased so that a current is produced by the photons as illustrated in Figure 4.17. The responsivity of the diode can be computed from the device quantum efficiency. The quantum efficiency, η, is defined as the average number of electrons produced by one photon. The photo current produced is

$$i_p = \eta q P_{in} \qquad (4.40)$$

where

q = the electronic charge, 1.602×10^{-19} coulombs per electron
η = the detector quantum efficiency, electrons per photon
P_{in} = the input optical power in watts

The photon flux is related to the optical power by

$$S_p = \frac{\lambda P}{hc} \qquad (4.41)$$

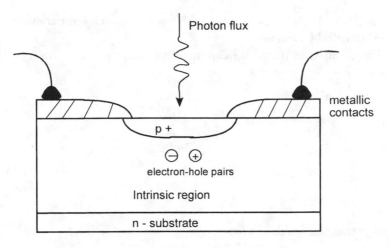

Figure 4.17 Photo diode.

where

h = Boltzmann's constant, 1.38×10^{-23} joules per K
c = speed of light, 3×10^8 meters per second
λ = wavelength in meters
P = the optical power in watts

Combining these equations, the detector responsivity in amperes per watt is obtained

$$R_d = 0.8064\eta\lambda \quad amp \, / \, watt \tag{4.42}$$

The quantum efficiency of the typical photo diode at 0.85 micron wavelength is about 0.7 with a responsivity of about 0.5 amperes per watt.

The avalanche photo diode is similar to the *p-i-n* diode except that a high back bias is used which can cause the electron-holes released by photons to have sufficient energy to release additional electrons. This produces a gain so that the responsivity is increased by the gain factor

$$R_{APD} = 0.8064\eta\lambda g \quad amp \, / \, watt \tag{4.43}$$

where

g = the avalanche gain

Gains of 100 to 1000 can be achieved although the gain is temperature dependent and statistically variable.

The creation of the photo current is a Poisson process with a mean and variance

$$i_{mean} = R_d P$$

$$< i_{ms}^2 > = 2qR_d PB \tag{4.44}$$

where

$B =$ the bandwidth in hertz

The signal-to-noise ratio, SNR, is

$$SNR = \frac{i_{mean}}{\sqrt{< i_{ms}^2 >}} = \sqrt{\frac{R_d P}{2qB}} \tag{4.45}$$

The optical power at which the signal-to-noise ratio is one (0 dB) can be computed as

$$P_{0dB} = \frac{2qB}{R_d} \tag{4.46}$$

The same limit can be computed for an ideal APD with gain. The photon limited SNR as a function of bandwidth is shown in Figure 4.18 for p-i-n and APD detectors with gains of 100, 200, and 500.

The photon limited noise is a fundamental limit to the photo diode performance. Photo diodes also exhibit leakage and dark currents. These currents also contribute shot noise to the received signal.

$$< i_{sn}^2 > = 2q(I_{dc} + I_{leak})B \tag{4.47}$$

The detector preamplifier noise must also be considered. The photo diode is a current source and the preamplifier is assumed to be implemented as a transimpedance amplifier as shown in Figure 4.19. The input of the preamplifier is a

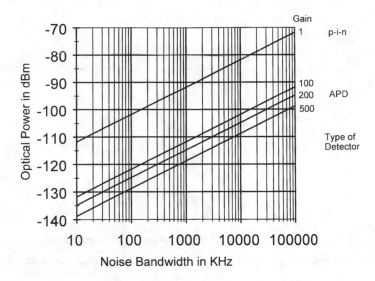

Figure 4.18 Photon limited noise power.

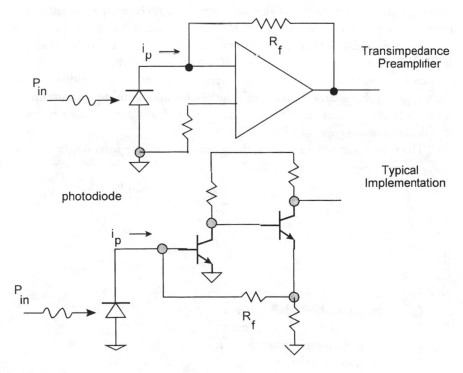

Figure 4.19 Photo diode preamplifier.

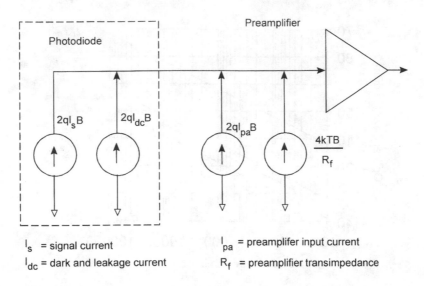

Figure 4.20　Photo diode detector noise model.

virtual ground so that the input signal current is balanced by the current in the feedback resistor. The amplifier gain is R_F volts per ampere (ohms) so the output voltage is the product of the signal current and R_F. A simplified noise model of the amplifier input is shown in Figure 4.20 which includes the shot noise due to the input transistor leakage current, the thermal noise due to the feedback resistor, the photo diode dark current, and the photo diode leakage current. Other noise components, such as 1/f noise, are neglected but are included in more complete published papers [6,7].

The total mean square noise at the receiver input is approximately

$$< i_n^2 > \approx 2q\left(i_s + I_{dc} + I_{leak} + I_{pa}\right)B + \frac{4kTB}{R_f} \qquad (4.48)$$

where

i　　= the mean photo current
I_{dc}　= the photo diode dark current
I_{leak}　= the photo diode leakage current
I_{pa}　= the preamplifier input dc current
R_f　= the transimpedance
B　　= the noise bandwidth

The detector signal-to-noise ratio is then

$$SNR = \frac{i_s}{\sqrt{<i_n^2>}} \qquad (4.49)$$

The maximum preamplifier transimpedance is a function of the bandwidth and cannot be chosen arbitrarily. The transimpedance amplifier bandwidth is determined by the pole formed by the input shunt capacitance and the equivalent input resistance. The input resistance is approximately the ratio of the transimpedance to the open loop gain. A commercially available preamplifier has a transresistance of 3.5 kilohms with an open loop gain of about 70 resulting in an equivalent input resistance of about 50 ohms. For an input capacitance of 7.5 pF, the bandwidth is about 400 MHz.

For a bipolar transistor amplifier, the input transistor has a base current, I_b, which has a shot noise component

$$<i_b^2> = 2qI_bB \qquad (4.50)$$

For a base current of 10 μA, the shot noise spectral density is 1.8 pA/$\sqrt{Hz}$. With a transimpedance of 3.5 kilohms, the RMS noise spectral density due to the transimpedance is

$$\sqrt{\frac{4kT}{R_f}} = 2.2\,pA\,/\,\sqrt{Hz} \qquad (4.51)$$

Thus, the total equivalent input noise current is about 4 pA/$\sqrt{Hz}$. The measured noise spectral density for this amplifier is 3.5 pA/$\sqrt{Hz}$ which is amazingly close to the calculated result considering some of the approximations. The computed noise over a 200 MHz bandwidth is about 56 μA while the measured value is 66 μA. The difference represents noise components such as 1/f noise which have been neglected.

The output signal-to-noise ratio (SNR) can be related to the input optical power via the detector responsivity, R_d. Combining (4.48) and (4.49) and including the detector responsivity, the SNR is

$$SNR = \frac{R_d P_{op}}{\sqrt{\left(2qR_d P_{op} + 2qI_t + \frac{4kT}{R_f}\right)B}} \qquad (4.52)$$

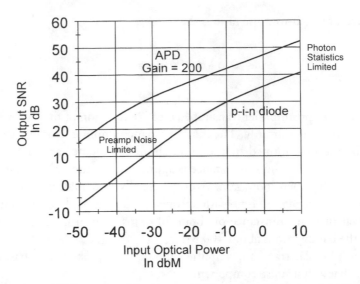

Figure 4.21 Typical SNR as a function of input optical power (100 MHz bandwidth).

where

$$I_t = I_{dc} + I_{leak} + I_{pa}$$
R_f = the detector responsivity in amperes per watt
P_{op} = the input optical power in watts

In the case of the avalanche photo diode, the avalanche gain is included in the responsivity and the excess noise introduced by the gain process is ignored. The SNR as a function of input power in dBm is plotted in Figure 4.21 for a bandwidth of 100 MHz, the preamplifier parameters listed previously and an avalanche gain of 200. The curves illustrate two ranges, one dominated by the preamplifier noise and the second region dominated by the photon statistics.

4.3 Design Charts

The system link analysis for cable systems is performed in the same manner as a wireless link. The medium transmission loss is computed, losses due to couplers are added, and the received power is determined. The minimum received power is determined according to the type of signaling used and the maximum acceptable bit error rate. The system design margin is the difference between the received power and the required power. Several nomographs are provided to quickly estimate key system parameters for fiber optic systems. Figure 4.22 computes the maximum transmission loss in terms of the distance and the loss coefficient (dB/km). The chart also computes the maximum distance for a given bandwidth and fiber bandwidth-distance parameter. With a fiber optic link, the systems designer must determine both the maximum distance for a given bandwidth and the maximum distance for a given loss allowance. This chart provides both calculations. The example shown in the chart computes the maximum transmission distance (1 km) for a fiber with a bandwidth of 100 MHz/km for a 100 MHz required bandwidth. In addition, the maximum distance (15 km) is computed assuming a maximum loss of 30 dB with a fiber loss of 2 dB/km. Of course, these examples can be computed mentally, but other choices of parameters may not be as easy.

Figure 4.23 is a chart to compute detector responsivity as a function of wavelength and quantum efficiency. With the responsivity, the detector current can be computed from the optical power. As an example, the responsivity is computed for a detector at 0.85 μm with a quantum efficiency of 0.7. The responsivity is slightly less than 0.5 A/W. The detector current for a −25 dBm input signal is about 1.5 μA. Two power and current scales are provided to cover the range from 0 to −60 dBm. The right hand power scale is used with the micro ampere range and the left hand scale is used with the nanoampere scale.

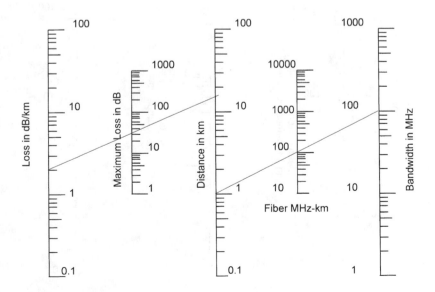

Figure 4.22 Fiber optic loss nomograph.

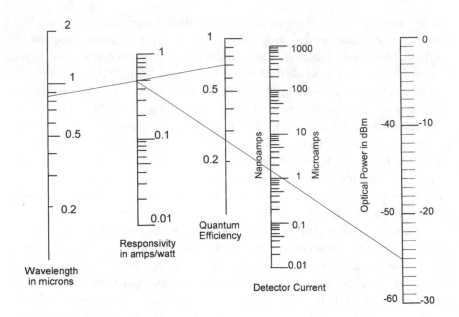

Figure 4.23 Optical detector sensitivity.

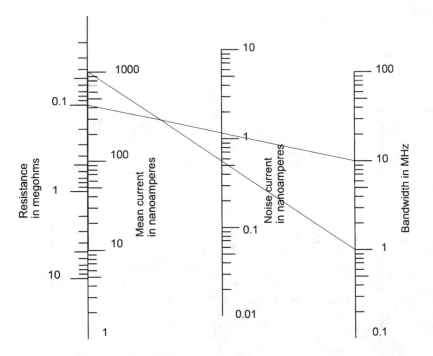

Figure 4.24 Noise current nomograph.

As an aid to computing the noise performance, Figure 4.24 computes the shot noise due to a given current and the Norton equivalent noise current for a resistor at a temperature of 300° kelvin. For example, the RMS noise current for a mean current of one μA (1000 nA) in a noise bandwidth of one MHz is 0.56 nA. The equivalent thermal noise current in a 100 kohm resistance for a bandwidth of 10 MHz is computed to be 1.2 μA. This chart in conjunction with the previous chart can be used to estimate the performance of the typical fiber optical detector.

References

[1]   Hayt, Wiliiam H., and Jack E. Kemmerly, *Engineering Circuit Analysis, Fourth Edition*, New York: McGraw-Hill Book Company, 1986.

[2]   Ramo, S., J. Whinnery, and T. Van Duzer, *Fields and Waves in Communication Electronics*, New York: John Wiley and Sons, 1965.

[3]   Heidemann, R., B. Wedding, and G. Veith, "10-GB/s Transmission and Beyond," *Proceedings of the IEEE*, Vol 81, November 1993.

[4]   Cochrane, P., and M. Brain, "Future Optical Fiber Transmission Technology and Networks," *IEEE Communications Magazine*, November 1988.

[5]   Scholl, F. W., et al, "Applications of Plastic Optical Fiber to Local Area Networks," *Fiber Optic Datacom and Computer Networks*, SPIE, Vol. 991, 1988.

[6]   Personick, S. D., "Receiver Design for Digital Fiber Optic Communication Systems, Part I and Part II," *Bell System Technical Journal*, Vol 52, July-August 1973.

[7]   Muoi, T. V., "Receiver Design for High Speed Optical Fiber Systems," *IEEE Journal of Lightwave Technology*, Vol. LT-2, 1984.

5

PCM Encoding and Modulation

Starting with a digital data stream, the PCM systems designer must decide how to encode the "ones" and "zeros" into symbols for transmission. Besides choosing the symbols, the designer must decide whether channel encoding and/or modulation are required. The encoding and modulation functions are shown in Figure 5.1. The input data stream is represented by a sequence of impulses with values $+1$ or -1 representing a data "one" or "zero," respectively. The channel encoder includes the functions of randomizing, encoding for error detection and correction, and multilevel encoding.

The symbol encoder converts the impulses into signaling waveforms appropriate to the transmission media. Often, the symbols must be translated in frequency to match the transmission channel frequency response. This translation is termed modulation and includes amplitude, phase, and frequency modulation. If the symbols are transmitted directly, the signaling is called "baseband." Baseband signaling normally implies the signal waveform energy is concentrated at low frequencies extending to dc. Some baseband codes do not have energy at dc but are considered baseband signals because of significant energy at low frequencies.

Radio frequency systems require modulation to translate the PCM signal to the desired frequency band. Multiple PCM data streams are often transmitted using frequency separation of the individual streams, a technique known as frequency division multiplex (FDM). Because channels are separated by frequency, the designer must be careful to reduce crosstalk from one channel to another. Radio systems are regulated by national and international emission

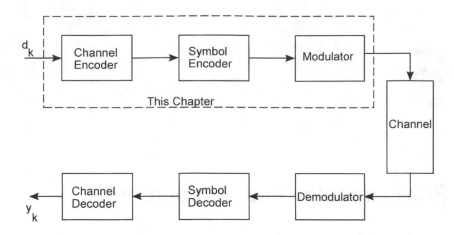

Figure 5.1 Encoding and modulation.

standards that define the allowable radiation outside the assigned channel. These regulations have a significant impact on the choice of modulation for the PCM stream.

Choosing the symbols, channel encoding, and modulation requires an understanding of the overall system performance requirements and the transmission link characteristics. In this chapter, we can only summarize some common PCM codes and modulation techniques with some hints as to their application. Once demodulation and detection are discussed in Chapter 6, the choice of codes and modulation will become clear and this material will serve as a design reference.

5.1 PCM Baseband Codes

The discussion of PCM codes will be separated into symbol codes and channel codes. The nomenclature "PCM code" can be ambiguous but is commonly used to denote the signal waveforms assigned to the input sequence. In this book, symbol codes will refer to the waveforms assigned to a given digital input. Channel codes will include the more conventional definition of error detecting and correcting codes. The simplest symbol code represents a "one" by one symbol waveform and a "zero" by a second waveform. More complex codes may have a set of rules that define the output signal waveform as a function of the input sequence. The system design task is to choose the symbol waveforms that optimize overall performance. In a nonbandlimited channel with additive, white Gaussian noise (AWGN), the bit error probability depends on the symbol energy-to-noise spectral density. For this type of channel, symbols with the greatest energy within the symbol duration and maximum level constraints should be chosen. The symbol with the greatest energy for a fixed duration and maximum level is a rectangular pulse and naturally defines the nonreturn-to-zero (NRZ) PCM code.

For a bandlimited channel, the rectangular symbol will be distorted creating intersymbol interference and degrading performance. For this channel, symbols that reduce intersymbol interference are desirable. Channels with poor or no low frequency response require symbols that can tolerate no dc response without excessive baseline wander. These are just two examples of the tradeoffs that must be considered.

5.1.1 Symbol Codes

5.1.1.1 Rectangular Symbols

Many PCM symbol codes have been proposed for different PCM applications. Standards organizations, such as the Inter-Range Instrumentation Group (IRIG), have defined recommended codes while other applications have gravitated to a set of generally accepted codes. The IRIG standard set of codes, shown in Figure 5.2, are commonly used for baseband PCM systems. The NRZ codes are best suited to channels with dc response, while the biphase codes (also

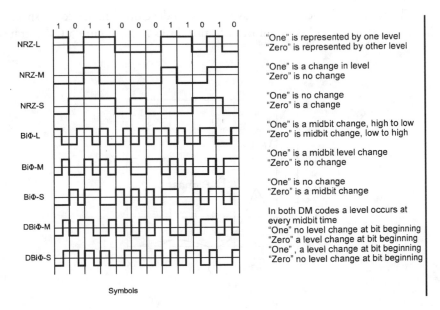

Symbols

Figure 5.2 IRIG standard PCM codes.

known as Manchester codes) are applicable to channels with no dc response or poor low frequency response. The power spectral density for the NRZ and biphase codes is shown in Figure 5.3. The NRZ codes concentrate the signal energy at low frequencies while the biphase spectral density peaks near a 0.75 bit rate.

The NRZ codes can have long sequences with no transitions creating baseline wander on channels with no dc response and the lack of transitions can cause symbol synchronization problems. The NRZ level codes are defined as absolute levels and are ambiguous on channels that can have polarity inversions. The mark and space codes eliminate this problem by encoding the data as level transitions. When NRZ is band limited, the intersymbol interference degrades the system performance. The degradation from theoretical performance is in the order of 1.5 to 2 dB when limited to a bandwidth equal to the bit rate. Accepting this degradation, NRZ conveys one bit per second per hertz bandwidth.

The biphase codes are designed for channels with no dc response and are commonly used for recording PCM. From the power spectral density, the biphase codes require a bandwidth of about twice the bit rate. In the search for PCM codes with no dc content and a spectral bandwidth less than biphase, recording applications have developed some specialized codes. Some of these codes are shown in Figure 5.4 with the power spectral density shown in Figure 5.5. The narrow power spectral density can be misleading and bandlimiting these codes can degrade the performance. The detection problem will be discussed in Chapter 6.

The telecommunications industry has also developed many codes specifically matched to the types of communication links used. Figure 5.6 illustrates

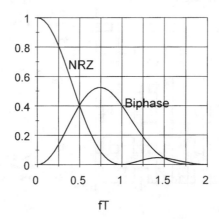

Figure 5.3 Power spectral density of NRZ and biphase codes.

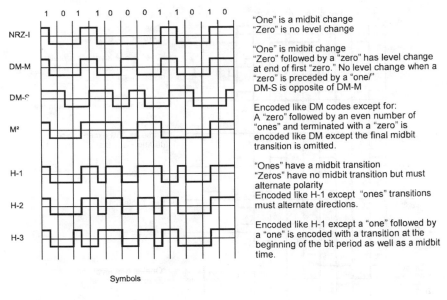

	1 0 1 1 0 0 0 1 1 0 1 0	
NRZ-I		"One" is a midbit change "Zero" is no level change
DM-M		"One" is midbit change "Zero" followed by a "zero" has level change at end of first "zero." No level change when a "zero" is preceded by a "one/" DM-S is opposite of DM-M
DM-S		Encoded like DM codes except for: A "zero" followed by an even number of "ones" and terminated with a "zero" is encoded like DM except the final midbit transition is omitted.
M²		
H-1		"Ones" have a midbit transition "Zeros" have no midbit transition but must alternate polarity Encoded like H-1 except "ones" transitions must alternate directions.
H-2		
H-3		Encoded like H-1 except a "one" followed by a "one" is encoded with a transition at the beginning of the bit period as well as a midbit time.

Symbols

Figure 5.4 PCM recording codes.

some of these codes. The coding rules for some more complex codes are summarized in Table 5.1

The unipolar codes are particularly appropriate for optical systems using LED or laser diode transmitters. The majority of these codes were developed for wireline transmission systems that are transformer- and capacitor-coupled with no dc response. The codes were also designed to maintain a minimum transi-

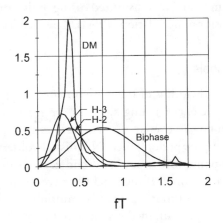

Figure 5.5 Power spectra of delay modulation, biphase, and Hedeman codes.

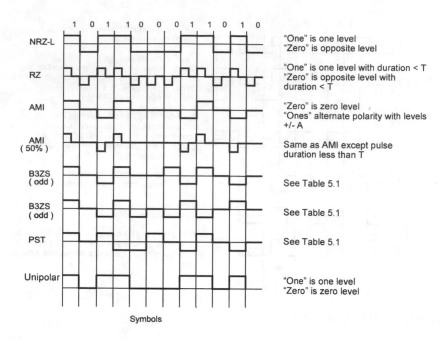

Figure 5.6 Telecommunication symbol codes.

tion density for network synchronization. The alternating polarity of the AMI codes provide a means of error detection for circuit monitoring. The power spectral density of the telecommunication codes is shown in Figure 5.7.

The NRZ and RZ codes can be extended to a multilevel symbol (also known as pulse amplitude modulation, PAM) conveying multiple bits per symbol. Digital data can also be transmitted using pulse duration modulation (PDM) and pulse position modulation (PPM). These codes are shown in Figure 5.8.

5.1.1.2 Nyquist Symbols

When the channel bandwidth is of the same order of magnitude as the bit rate, or less, bandwidth efficient signaling symbols are required. Any filtering will cause rectangular symbols to overlap into adjacent symbol periods (intersymbol interference) degrading the performance. Nyquist [1] observed that an ideal filter with constant response in the passband and zero response in the stopband has a $\sin(x)/x$ impulse response. If the ideal filter cutoff frequency is at one half the impulse rate, the filter output is zero at all multiples of the impulse period as shown in Figure 5.9. If the combined response of the symbol encoder, the communication link and the detection filter can approximate the ideal response,

Table 5.1
Coding Rules

B3ZS (Binary 3 zeros substitution)	The basic AMI line code is used. In the B3ZS code, every string of three zeros is replaced with either 00V or B0V, where V is a bipolar violation and B normal bipolar transition. If an odd number of a "ones" has been transmitted since the last substitution, 00V is substituted for the three zeros, otherwise, B0V is substituted.

B6ZS
(Binary 6 zeros substitution)
	The basic AMI line code is used. A string of 6 zeros is substituted with the following sequence depending on the polarity of the previous pulse:
	
	Previous pulse Substitution sequence
	  − 0−+0+−
	  + 0+−0−+

HDB3 (high density bipolar 3) (CCITT)	The basic AMI line code is used. Strings of 4 zeros are substituted according to the following rules:	
	Previous pulse	Number of "ones" since last sub.
		Odd Even
	−	000− +00+
	+	000+ −00−

PST (pair selected ternary)	The AMI line code is used but the transmitted symbols are selected according to the following:	
	Input pair Mode 1	Mode 2
	00 −+	−+
	01 0+	0−
	10 +0	−0
	11 +−	+−
	A mode is used until a single pulse is transmitted then the other mode is used.	

signaling can be achieved at 2 symbols per second per hertz with no intersymbol interference. Approximating the ideal response is a tall order even in this era of digital signal processors. Even if the ideal response could be approximated, the sampling at the receiver must be perfect.

If some excess bandwidth is allowed, zero intersymbol interference can still be achieved with practically realizable filters. In order to have zero intersymbol interference, all that is required is that the frequency response be symmetrical about the 0.5 response frequency, as illustrated in Figure 5.10. The amount of excess bandwidth ranges from 0% to 100%. The 100% excess bandwidth

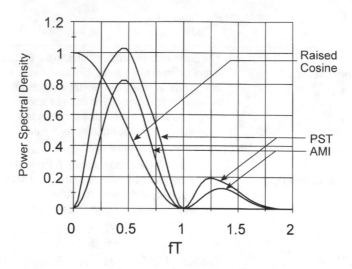

Figure 5.7 Power spectral density of some telecommunication codes.

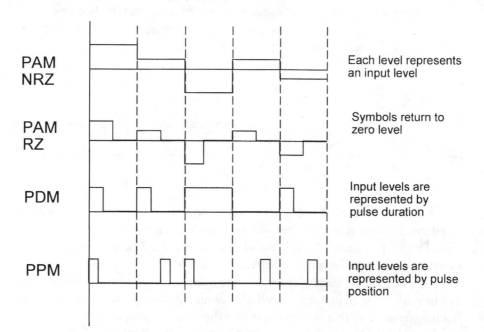

Figure 5.8 PAM, PDM, and PPM codes.

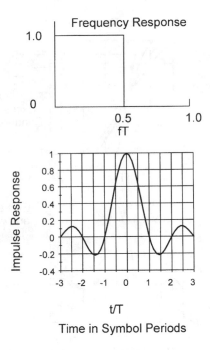

Figure 5.9 Ideal Nyquist filter.

case is called a raised cosine filter. While the raised cosine response only achieves a 1 symbol per second per hertz bandwidth efficiency, the response is much more tolerant to bandlimiting than the NRZ symbol.

With a band limited channel, the design goal is to have an end-to-end Nyquist response. This can be realized using a Nyquist symbol line coding with a sharp cutoff detection filter or by splitting the Nyquist response between the transmitter and receiver, as shown in Figure 5.11. The latter approach uses a square root filter to generate the transmitted symbols. The impulse response of the square root filter (also called the half Nyquist filter) is calculated using the inverse fast frequency transform (FFT) of the frequency response. The symbol encoder can be synthesized using digital logic and read-only memory [2].

Nyquist assumed the symbols to be independent. If this assumption is removed, a class of symbols known as partial response signaling can be constructed which maintain zero interference while increasing the signaling efficiency. The idea behind partial response signaling is to introduce a controlled intersymbol interference that reduces the signal bandwidth. Since the intersymbol interference is known at the receiver, it can be removed before the symbol decision is made. The first partial response signaling technique (called

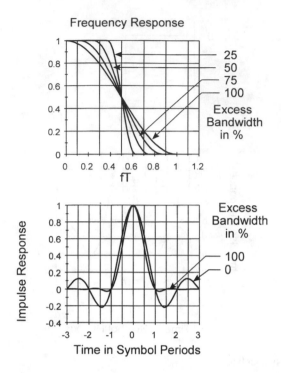

Figure 5.10 Nyquist signaling with excess bandwidth.

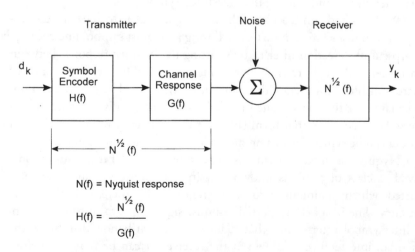

Figure 5.11 Nyquist signaling using half Nyquist filters.

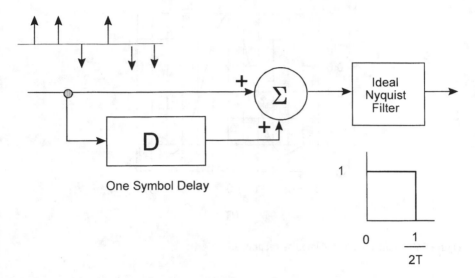

Figure 5.12 Duobinary signaling.

duobinary) was introduced by Lender [3] using a correlation between successive bits as shown in Figure 5.12. The impulse response of the Nyquist filter is a sinc(x) function so the impulse response of the duobinary signal is

$$h(t) = \frac{\pi}{4}\left[\text{sinc}\left(\pi\left(fT + \frac{1}{2}\right)\right) + \text{sinc}\left(\pi\left(fT - \frac{1}{2}\right)\right)\right] \qquad (5.1)$$

where

$$\text{sinc}(x) = \frac{\sin(x)}{x}$$

The impulse response is shown in Figure 5.13 illustrating the correlation between successive symbols with zero intersymbol interference elsewhere.

The one symbol delay and adder used to create the correlation between symbols can be considered to be a digital prefilter defined as

$$f(D) = 1 + D \qquad (5.2)$$

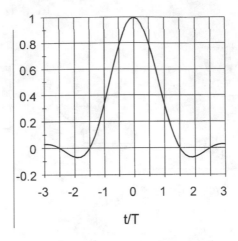

Figure 5.13 Duobinary signal impulse response.

where

D is a delay operator, e^{-sT}

The frequency response of the prefilter is

$$F(s) = 1 + e^{-sT} \qquad (5.3)$$

If the prefilter frequency response is evaluated, the magnitude of the response is

$$\left| F(j\omega) \right| = 2\cos(\pi fT) \qquad (5.4)$$

When combined with the Nyquist filter, the duobinary frequency response is shown in Figure 5.14. The bandwidth is limited to less than 0.5 the symbol rate with a signaling efficiency of 2 bits per second per hertz.

The duobinary signal is a three level signal taking values, −2, 0, and +2. The signal must always make the transition through zero from either −2 or +2. Thus, the zero level is always associated with a data transition while the −2 and +2 levels represent no transition. This suggests one means for detecting the signal at the receiver.

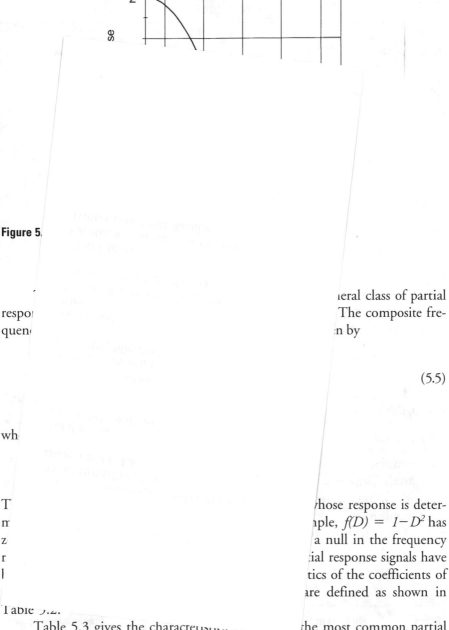

Figure 5.

ieral class of partial
respo⟩ The composite fre-
quen⟨ ⟩n by

$$\tag{5.5}$$

wh⟩

T ⟩hose response is deter-
n ⟩ple, $f(D) = 1 - D^2$ has
z ⟩a null in the frequency
r ⟩ial response signals have
| ⟩tics of the coefficients of
⟩are defined as shown in
Table 5.2.

Table 5.3 gives the characteristics the most common partial response signals. The partial response signals all produce multilevel signals which can be undesirable on links with nonlinear distortion. The higher the order the polynomial, the greater the number of levels and the less useful the signal.

Table 5.2
Partial Response Classes

Partial Response Class	Definition	Example Polynomials
1	All coefficients have value, $+1$	$1+D, 1+D+D^2$
2	The polynomials have an odd number of coefficients with the coefficients weighted as a symmetric triangle	$1+2D+D^2$, $1+2D+3D^2+4D^3+3D^4+2D^5+D^6$
3	The coefficients start with amplitude, $n-1$, decrease by one and alternate signs between successive symbols	$2+D-D^2$, $4+3D-3D^2+2D^3-2D^4+D^5-D^6$
4	Similar to class 2 but the coefficients are weighted with a positive and negative triangle weighting	$1+0D-D^2$, $1+2D+D^2+0D^3-D^4-2D^5-D^6$

5.1.2 Channel Codes

Channel coding is used primarily to combat symbol errors introduced in the communication link. In addition to error detection and correction, randomizing and group coding may also be used to modify the transmitted signal transition density to improve the synchronization performance.

5.1.2.1 Randomizing

The communications system designer has no control over the content of the data stream presented to the PCM system. As a result, the data can have long sequences of bits of the same level. If this data is encoded using NRZ codes, the lack of transitions makes it very difficult for the receiver to acquire or maintain symbol synchronization. In some cases, the designer may have control over the

Table 5.3
Some Common Partial Response Signals

Class	Number of Signal Levels	Prefilter	Frequency Response
1	3	$1+D$	$2\cos(\pi fT)$
2	5	$1+2D+D^2$	$1+\cos(\pi fT)$
4	3	$1-D^2$	$2\sin(2\pi fT)$

data format and can insure a minimum data transition density by designing the word or frame format.

A more general approach that requires no knowledge of the data sequence uses a pseudo random (PR) sequence to "randomize" the data stream prior to the symbol encoder. A PR sequence is a binary sequence (1 or 0) which is added, modulo-2, to the binary data stream. Long sequences of "ones" or "zeros" are broken up and the average transition density approaches 0.5. The PR sequence is known at the receiver and can be removed from the detected data stream by adding, modulo-2, a replica of the randomized output to the received data stream.

The PR sequence is generated using a linear sequential shift register with exclusive-OR gates used to perform the modulo-2 addition. The same shift register is used at the receiving end to remove the randomizing sequence as shown in Figure 5.15. The derandomizer is self-synchronized with the randomizer. With no channel errors, the same sequence enters the shift register at the receiver and the transmitter. Therefore, the derandomizer output sequence is the same sequence added to the original data stream and adding it modulo-2 to the received stream will restore the original data. If a single error occurs within the length of the shift register, more than one error will be produced. The error multiplication is equal to the number of feedback taps on the register, typically, two. Randomizing the data will typically double the received error rate, a small penalty in equivalent signal-to-noise ratio.

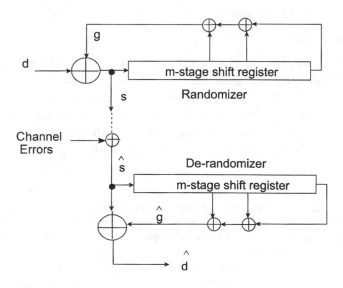

Figure 5.15 Data randomizer.

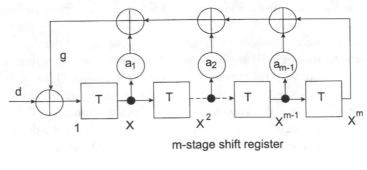

$$g(X) = 1 + a_1 X + a_2 X^2 + \ldots + a_{m-1} X^{m-1} + a_m X^m$$

Figure 5.16 Polynomial representation of PR sequence generator.

For a shift register of length, L, the PR sequence length is $2^L - 1$. The PR sequence generator design [4,5] is based on primitive polynomials that define the tap connections of the shift register. If the outputs of the shift register stages are considered to be powers of a polynomial, $g(x)$, the taps of the polynomial to be used for feedback are given by the nonzero coefficients as shown in Figure 5.16. Two of the most common randomizers use 11 and 15 stage generators with tap connections given in Table 5.4.

There are several primitive polynomials for a given shift register length and the tap connections shown for the 11 and 15 stage registers have the minimum number of taps and, consequently, the least error multiplication.

5.1.2.2 Group Codes

Group coding provides another approach to controlling the sequence transition density. The idea of group coding is to take a group of bits and map them into an output group with one additional bit. The extended group can be designed to have a prescribed number of bit transitions. As an example, a 4-bit group of bits represents 16 different words, appending an additional bit increases the number of possible words to 32, and the 16 codes with the best transition densi-

Table 5.4
PR Sequence Generator Tap Connections

Shift Register Length	Sequence Length	Tap Connection Polynomial
11	2047	$x^{11} + x^2 + 1, x^{11} + x^9 + 1$
15	32767	$x^{15} + x + 1, x^{15} + x^{14} + 1$

ties can be output. Since there are codes that cannot be produced by legitimate data words these can be used for synchronization or error monitoring.

Group coding requires segmenting the input data stream into groups of bits at the transmitter and extracting encoded groups at the receiver. The design of the group codes is largely brute force and the method is limited to relatively short group lengths. Some of the symbol codes such as binary n-zeros substitution (BNZS) and the 3PM code can be described as specialized group codes.

5.1.2.3   Error Correcting Codes

The major thrust of channel coding is the detection and correction of channel errors. The simplest model of the PCM communication link with error detection and correction is shown in Figure 5.17. The binary symmetric channel (BSC) assumes a transmitted "one" and is detected as a "one" at the receiver with probability, $1-p$. Errors are made with probability, p. The encoder introduces redundancy to correct channel errors. There is a wealth of literature dealing with error correcting codes with a level of detail ranging from elementary to highly sophisticated [4–10]. The PCM systems engineer is primarily concerned with the decoded performance of the system for a given coding method that will be discussed in Chapter 6. In this section, the general characteristics of the coding methods will be summarized.

There are three generic error correcting coding methods:

- Block codes;
- Convolutional codes;
- Concatenated codes.

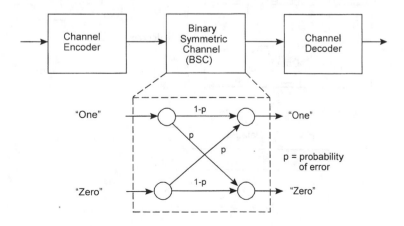

Figure 5.17   Error correction channel model.

In addition to these methods, several recent methods such as trellis code modulation (TCM) and turbo codes use the basic codes in combination or with modulation. All of these coding methods add redundancy to the data stream that is used to detect and correct transmission errors. The codes are described by the code rate, k/n, the ratio of the information bits to the total transmitted bits.

Block codes take k information bits and append $N-k$ bits (parity) to form the N bit word for transmission as shown in Figure 5.18. A k-bit code word can be visualized as a vector spanning a space with 2^k possible points. Each distinct code word occupies one point in the space. Adding the parity bits expands the code space from 2^k to 2^N possible signal locations. The performance of the coding depends on the distance between code vectors which is measured by the Hamming distance. The Hamming distance is the number of bit positions in which two code words differ. For two code words to be distinct, they must be separated by a Hamming distance of at least one. If one parity bit is added, the Hamming distance is increased by one so a single error produces a vector that is still separated from the code words by a minimum distance of one. Thus, a single error can be detected with a Hamming distance of two. If the distance is increased to three, single errors can be detected and corrected. This relationship can be generalized. To correct t errors, the minimum Hamming distance must be

$$d \geq 2t + 1 \tag{5.6}$$

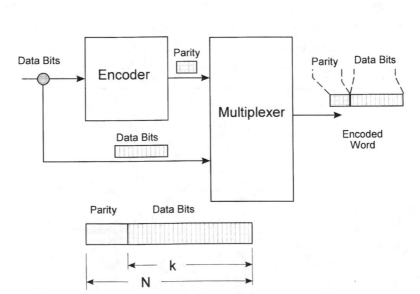

Figure 5.18 Block encoder.

Similarly, to detect v errors, the distance must be

$$d \geq v + 1 \tag{5.7}$$

The design of block codes has been a search for codes that have as great a separation as possible within the 2^N space. Although the basic idea is simple enough, finding systematic methods to design good block codes is extremely difficult.

The cyclic codes discovered by Bose, Chaudhuri, and Hocquenghem [11] (commonly called BCH codes) have a systematic design procedure and can be implemented with linear shift register circuits. Table 5.5 lists the BCH codes that correct up to four errors for lengths to 511 bits.

The number of parity bits for a BCH code that can correct t errors is

$$r \leq mt$$

$$N = 2^m - 1 \tag{5.8}$$

where

r = the number of parity bits
N = the code length

If the error correction efficiency is defined as the ratio of the correctable errors to the code length

$$R_{err} = \frac{t}{N} \tag{5.9}$$

Table 5.5
BCH Codes

Length	Information Bits	Code Rate	Correctable Errors
7	4	0.57	1
15	11, 7, 5	0.733, 0.466, 0.333	1, 2, 3
31	26, 21, 16	0.838, 0.677, 0.516	1, 2, 3
63	57, 51, 45, 39	0.905, 0.810, 0.714, 0.619	1, 2, 3, 4
127	120, 113, 106, 99	0.944, 0.889, 0.835, 0.779	1, 2, 3, 4
255	247, 239, 231, 223	0.968, 0.937, 0.905, 0.874	1, 2, 3, 4
511	502, 493, 484, 475	0.982, 0.965, 0.947, 0.929	1, 2, 3, 4

The code rate, k/N, is then related to the error correction efficiency

$$R_{code} = \frac{k}{N} = \frac{N - mt}{N} = 1 - mR_{err} \qquad (5.10)$$

For a given error correction efficiency, the code rate decreases as the logarithm of the code length. Thus, the longer the BCH code, the less efficient it is at error correction. Despite all efforts at designing good block codes, all known block codes have this problem.

The BCH code uses binary symbols but nonbinary codes can also be developed. The Reed-Solomon (RS) codes are the best example of the nonbinary cyclic code. The RS code word is shown in Figure 5.19 with the binary symbols replaced by q-bit symbols. The RS codes are a special case of nonbinary codes with a code length of $2^q - 1$ and $2e$ parity symbols (e are the correctable symbols.) If the code can correct e symbols and each symbol is q bits long, then bit error bursts of qe can be corrected. Thus, a single error correcting code with 8 bit symbols can correct up to 8 bit error bursts. Applications of RS codes are described in [4].

The next category of error correcting codes is the convolutional code. Where the block code groups the information bits in a block with appended parity bits, the convolutional code interleaves parity bits in the serial data stream. The code rate is the ratio of the information bits to the parity bits and convolutional codes can only have rational fractional rates of the form, b/n. The convolutional code is best described by example. Figure 5.20 shows the encoder for a 1/2 rate code. In this example, the data stream is clocked into a shift regis-

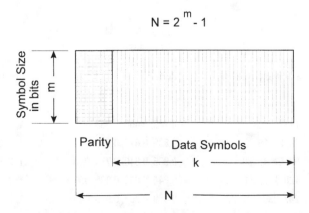

Figure 5.19   Reed-Solomon codes.

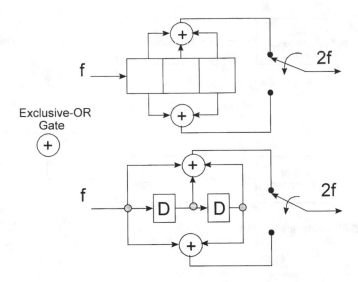

Figure 5.20 Two representations of a nonsystematic half rate convolutional code.

ter at the bit rate. The combinations of the taps used to form the outputs are defined by the code generator polynomials. At each bit time, one bit is taken from the upper output and one bit is taken from the lower output with the output rate equal to twice the bit rate. When the input data is used as one output, the code is called "systematic." In this case, a 1/2 rate code would have alternating data and parity in the output stream.

In addition to the code rate, the convolutional code is characterized by the constraint length which is usually defined as the length of the shift register in the encoder. The constraint length represents the length over which an isolated error will affect the code before it is flushed out of the register. The constraint length divided by the code rate is analogous to the length of a block code. A 1/2 rate, constraint length 11 has an effective length of 22 bits. Most convolutional codes have been found by exhaustive computer search. A tabulation of some common convolutional codes is given in [12].

The idea of combining coders is more than 30 years old and is a powerful method for combating channel errors. A concatenated coder is one that uses two coders, an inner and an outer coder as shown in Figure 5.21. The probability of error can be made to decrease exponentially with code length

$$P_e \leq e^{-NE} \tag{5.11}$$

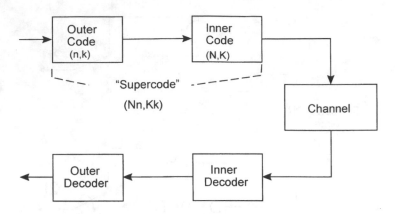

Figure 5.21 Concatenated coding.

where

N is the block length

E is a lower bound error exponent that depends on code rate

Theoretically, the longer the block length, the lower the error rate. Unfortunately, longer block lengths require more complex decoding. Further, known block code error correcting efficiency decreases with increasing length. Forney [13] suggested a way out of the complexity problem by concatenating two codes. The concatenated code has an effective block length of $N_{inner} \times N_{outer}$ and a code rate of $R_{inner} \times R_{outer}$. Small codes can be used for the inner and outer codes and still achieve a long block length and low overall error rate. Concatenation is also an effective means for combating burst errors.

The error correction process consists of two steps, detecting an error and finding its location. In a binary code, finding the error location is tantamount to correcting the error since the bit at the error location only needs to be inverted. A nonbinary code, such as an RS code, must also compute the error value in addition to the location to correct the error. If the inner code corrects some errors and passes the uncorrected errors to the outer code, the decoder complexity can be greatly reduced. As an even simpler example, consider an array of data in which each row and column is encoded with a block code. The row code will be called the inner code and the column code, the outer code as shown in Figure 5.22. If the inner code only detects errors, it can tell the outer code the row, or rows, containing errors. If the outer code detects columns in error, the row error information immediately identifies the bits in error. Thus, each decoder is a simple error detector and bursts of errors can be corrected as well as random errors.

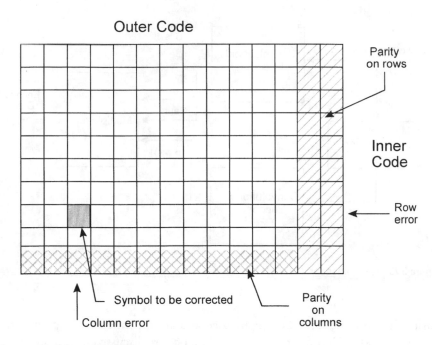

Figure 5.22 A concatenated code example.

Concatenated codes are widely used on communication links requiring performance near the theoretical bounds such as deep space probe links. High rate, high density recorders also use this technique to combat burst errors common in recording applications.

TCM, introduced in 1987, combines convolutional coding with quadrature amplitude modulation (QAM) to improve the performance of highly band limited systems. In one implementation, a 1/2 rate convolutional encoder is applied to one bit of a N-bit symbol creating an $N + 1$ bit symbol. The two bits from the convolutional encoder partion the symbols into four sets. The distance between signal points in the four sets is increased from two to four and the net gain in performance is about 3 dB. TCM has been implemented in high rate telephone line modems.

Recently, a new coding development has burst onto the scene accompanied with both skepticism and praise. The coding method [14,15] has followed the trend of modern car makers and is called Turbo coding. Amazingly, the Turbo code performance approaches the Shannon coding limit within 0.7 dB. The basic configuration of the Turbo encoder is shown in Figure 5.23. Two encoders are used with the first, encoding the input data stream directly. The input data stream is interleaved and encoded using a second encoder. Only the

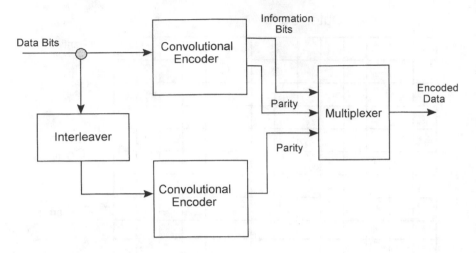

Figure 5.23 Turbo encoder.

parity bits from the second coder are used and are appended to the parity from the first encoder. Decoding and the performance are discussed in Chapter 6.

5.2 Modulation and Multiplexing

Many communication channels will not support baseband signal transmission and some type of modulation is required to translate the signal to a usable band. Amplitude modulation offers a direct means for translating a baseband signal to a bandwidth centered on the carrier frequency. Phase and frequency modulation of a carrier offer alternative means for translating the baseband signal. Spread spectrum modulation addresses a different goal, combating channel interference or intentional jamming.

This book deals only with digital modulation (discrete level) modulation methods. Historically, these methods have been described using the term "keying," frequency shift keying (FSK), phase shift keying (PSK), and so on. These terms remain in common use and will be used here.

Before considering modulation methods for PCM, a few words about multiplexing is in order. Most PCM systems are designed to convey multiple sets of data. A telecommunication system transmits multiple voice channels; a telemetry system transmits multiple data channels, and so forth. Three types of multiplexing are commonly used, time-division-multiplexing (TDM), frequency-division-multiplexing (FDM) and code-division-multiplexing (CDM). The first two types are the most dominant, the latter type (CDM) is important

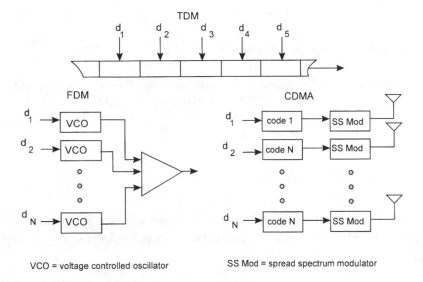

VCO = voltage controlled oscillator   SS Mod = spread spectrum modulator

Figure 5.24 Multiplexing methods.

in wireless telecommunication systems. The three types of multiplexing are illustrated in Figure 5.24. Systems can use combinations of these multiplexes, for example, TDM subcarriers in an FDM.

Time-division-multiplexing has become the dominant PCM multiplexing method. Multiple data channels are encoded and inserted into preassigned time slots of a composite serial data stream. Sometimes, time slots may also be multiplexed by several data streams. This is called subcommutation. Several main stream time slots may be used by a single data stream. This is called supercommutation. Some typical TDM formats are discussed in [4].

Frequency-division-multiplexing (FDM) assigns a different frequency carrier for each primary data stream separating channels in the frequency spectrum. Broadcast television and satellite communications are typical examples. The individual carriers are often called subcarriers and many subcarriers are modulated with TDM.

Code-division-multiplexing is a method of separating individual data streams by encoding the streams using an orthogonal code. The orthogonal code has the property that

$$\int_0^T c_i c_j \, dt = 1 \quad \text{for } i = j$$

$$= 0 \quad \text{for } i \neq j \qquad (5.12)$$

where

     c_i is the code for the i-th data

By appropriate signal processing, the individual signals can be separated using the orthogonality property. CDM has established an important application in spread spectrum wireless systems.

5.2.1   Amplitude Modulation

Amplitude modulation is probably the oldest, and simplest, method used. The modulation is represented as the product of the baseband signal and the carrier signal

$$y(t) = m(t)c(t) \tag{5.13}$$

where

     $m(t)$ is the baseband PCM signal
     $c(t)$ is the carrier signal

The frequency spectrum of the modulated output is the Fourier transform of the product of the baseband and carrier functions. The Fourier transform of a product is the convolution of the transforms of the individual functions [16].

$$Y(\omega) = \int_{-\infty}^{+\infty} M(\Omega)C(\Omega - \omega)d\Omega \tag{5.14}$$

The modulating carrier can be assumed to be a sine function, a cosine function, or a linear combination of the two. The Fourier transform of the sine and cosine functions are

$$\cos(\omega_0 t) \Leftrightarrow \pi\left[\delta(\omega + \omega_0) + \delta(\omega - \omega_0)\right]$$

$$\sin(\omega_0 t) \Leftrightarrow j\pi\left[\delta(\omega + \omega_0) - \delta(\omega - \omega_0)\right] \tag{5.15}$$

where

$\Leftrightarrow$ denotes the Fourier transform relationship

The convolution process can be described as the computation of the overlapping area of the product of the two functions as one is slid past the other. Both the sine and cosine functions have transforms which are impulse functions so the convolution of these functions with the baseband signal spectrum translates the baseband signal spectrum to a band around the carrier frequency. This process is illustrated in Figure 5.25 for a cosine carrier. In practical systems, the translation from baseband to carrier frequencies is accomplished in several stages, translating the baseband signal first to one, or more, intermediate frequencies (IF) before reaching the final frequency. At each IF stage, filtering must be used to remove undesired sidebands. The spectral occupancy of the modulated carrier can be directly computed from the baseband signal spectrum. If the baseband signal is 40 dB down relative to dc at frequency, f_0, the carrier signal (frequency, f_c) will be 40 dB down at $f_c + f_0$ and $f_c - f_0$.

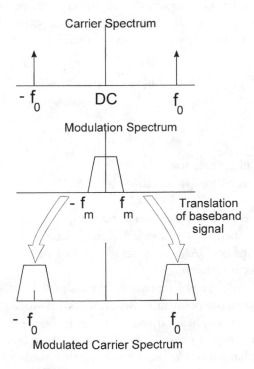

Carrier Spectrum

$- f_0$ DC f_0

Modulation Spectrum

$- f_m$ f_m Translation of baseband signal

$- f_0$ f_0

Modulated Carrier Spectrum

Figure 5.25 Amplitude modulation process.

If the modulating signal is a unipolar, return-to-zero (RZ) symbol, the carrier modulation is called on-off-keying (OOK). Direct modulation of LEDs and laser diodes is a common application using OOK. The remote control units used for consumer electronics is a 32 to 40 kHz pulse carrier modulated by pulses with one duration representing a "one" and a second duration representing a "zero" (pulse duration modulation).

Since OOK depends on the signal amplitude, it is sensitive to any nonlinearities in the signal path. Apart from introducing signal distortion, the nonlinearities create spurious spectral components which can exceed the authorized channel bandwidth.

For a bipolar NRZ modulating signal, the input takes on one of two values, $+A$ or $-A$. The carrier is either unchanged or reversed in polarity. This modulation is equivalent to a 180 degree phase modulation, or PSK. If the transition from one level to another is nearly instantaneous, the carrier has a constant envelope. The constant envelope property is very desirable since it is insensitive to nonlinear distortion. When the transition is not instantaneous, the amplitude is no longer constant and nonlinearities can create spurious spectral components called "spectral regrowth."

Pulse amplitude modulation (PAM) uses multilevel symbols, representing multiple bits per symbol, to modulate the carrier. The spectral efficiency (bits per second per hertz) can be doubled if an orthogonal carrier is modulated by a second data stream and added to the in-phase carrier.

$$y(t) = m_I \cos(\omega_0 t) + m_Q \sin(\omega_0 t) \qquad (5.16)$$

where

   m_I is the in-phase data stream
   m_Q is the quadrature data stream

This modulation is cleverly called quadrature amplitude modulation, or QAM. The obvious implementation of the QAM modulator is shown in Figure 5.26. There are variations on QAM using staggered data streams so that the transitions do not all occur simultaneously.

The QAM signal can be represented by a signal "constellation" showing the location of the signal point in a two-dimensional space whose axes are the in-phase and quadrature-phase carriers as shown in Figure 5.27. When the two streams are binary, the in-phase and quadrature-phase components are both PSK and the resultant modulation is equivalent to quadraphase (four phases) PSK, or QPSK. If the amplitudes of the modulating signals are chosen such that

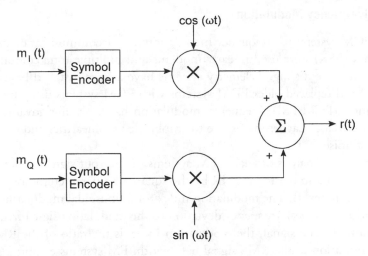

Figure 5.26 QAM modulator design.

the resultant phasor amplitude is constant ($m_I^2 + m_Q^2 = $ a constant), a multi-phase shift modulation, or M-ary, PSK is generated. By choosing the amplitude pairs, arbitrary signal constellations can be designed. If a constraint is placed on mean square carrier power, optimum signal constellations can be chosen which maximize the distance between signal points.

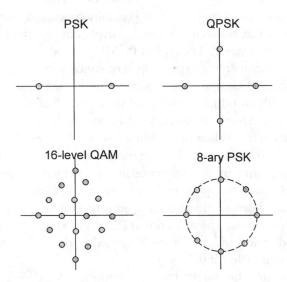

Figure 5.27 QAM signal constellations.

5.2.2 Frequency Modulation

Many PCM systems use frequency modulation. In telecommunication systems, two tones (FSK) were used in early modems and multiple tone combinations are used in tone dialing. Telemetry systems have a large installed base of FM receivers and replaced older FDM systems with PCM systems using frequency modulation (PCM/FM). Frequency modulation has the major advantage of a constant envelope carrier insensitive to amplitude nonlinearities and a good rejection of noise.

From the analysis of analog FM systems, the output signal-to-noise ratio (SNR) is related to the input SNR by the square of a parameter known as the modulation index, β. The modulation index for a sinusoidal modulating signal is the ratio of the peak frequency deviation to the modulating signal frequency. For a more general signal, the modulation index is the ratio of the RMS frequency deviation to the RMS signal bandwidth. FM systems exhibit a strong, nonlinear threshold above which the output SNR exceeds the input SNR and below which the system output SNR deteriorates rapidly. With a binary modulation, the modulation index is logically defined as the ratio of the peak-to-peak frequency deviation to the bit rate.

5.2.2.1 Frequency Shift Keying (FSK) and PCM/FM

As PCM/FM systems were developed, designers searched for the optimum modulation index to minimize system bit error rate. This search included both analytic and experimental studies. It was generally concluded that the optimum modulation index was between 0.7 and 1.5 depending on the type of receiver and premodulation filtering. The typical PCM/FM modulator is shown in Figure 5.28. The carrier frequency spectrum is of concern in an FM system where the spectral occupancy is limited. Frequency modulation by a single tone produces a signal with an infinite number of sidebands. Bandlimiting the carrier signal will distort the received signal. With binary modulation the carrier spectrum is more complex. When the modulation index is very large, implying a large number of carrier cycles per bit, the spectrum should approach that of two discrete tones separated by the difference in tone frequencies. As the modulation index is decreased, the spectrum between the tone frequencies will be filled in and at low modulation indexes, the spectrum should approach the $\text{sinc}(x)$ spectrum of a single tone. The spectrum of PCM/FM is sketched in Figure 5.29 for several modulation indexes. Interestingly, the spectrum is most compact near the "optimum" value of 0.7.

When discrete tones are used in FSK, optimum detection theories suggest that the tones should be orthogonal so that there is no crosstalk from one detec-

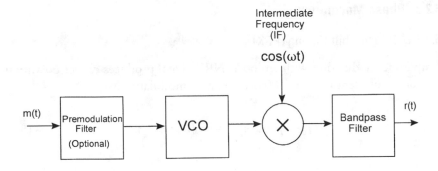

Figure 5.28 PCM/FM modulator.

tion filter into the other. The detection performance depends on whether coherent or incoherent detection can be used. Coherent detection requires the receiver to have knowledge of the phase and frequency of the carrier while incoherent detection uses envelope detection for separated tones or FM discriminator detection for continuous phase modulation. These factors are considered in Chapter 6.

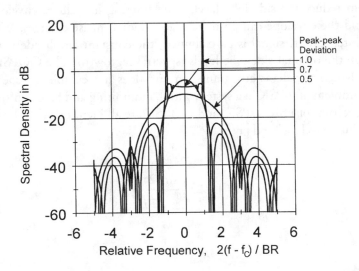

Figure 5.29 PCM/FM spectrum for several values of modulation index.

5.2.3 Phase Modulation

5.2.3.1 Phase Shift Keying (PSK)

Amplitude modulation with a bipolar NRZ signal produces carrier polarity inversions equivalent to $+/-$ 180 degree phase modulation.

$$y_k(t) = m_k \cos(\omega_0 t) = \cos(\omega_0 t + a_k \pi) \tag{5.17}$$

where

$m_k = +1 \text{ or } -1$
$a_k = 0 \text{ if } m_k = +1$
$a_k = 1 \text{ if } m_k = -1$

The 180-degree phase modulation is PSK. PSK is a constant envelope modulation with antipodal (opposite) signaling symbols. In the signal constellation, the symbols are diametrically opposite giving the best possible performance for a two level signal. By modulating an orthogonal carrier with a second data stream using a QAM modulator, quadraphase PSK, or QPSK, is formed. Other variations of QPSK include delaying one data stream by one half symbol to limit phase changes to zero or $+/-90$ degrees at data transitions. This modulation either is called staggered QPSK or offset QPSK (OQPSK).

Although ideal QPSK has a constant envelope, QPSK is frequently band limited to reduce the side lobe levels. The filtering introduces envelope variations and the envelope can go to zero at π radian phase changes. When the band limited QPSK signal is hard limited, the constant amplitude is restored but so are the original side lobe levels (spectral regrowth). The OQPSK phase cannot change π radians at a transition and does not exhibit the spectral regrowth problem of QPSK in a system with bandlimiting and hard limiting. The power spectrum of PSK, QPSK, and OQPSK are all identical with the spectral density shifted to baseband given by

$$P_{PSK}(f) = 2T \left(\frac{\sin(2\pi fT)}{2\pi fT} \right)^2 \tag{5.18}$$

where

$T = $ the symbol period

PSK can be extended to more than four phases and the general case is called M-ary PSK for M-levels. More than 8-phase PSK systems are less efficient than other modulations and not often used.

5.2.3.2 Minimum Shift Keying (MSK) and Continuous Phase Modulation (CPM)

In the course of investigating FSK signals with low modulation indexes ($\beta < 1$), the idea that phase knowledge of the FSK signal should aid in the signal detection was introduced. A modulation known as minimum shift keying (MSK) was developed which can be thought of either as a phase modulation method or as a special form of FSK with a modulation index of 0.5 and a continuous phase. Because of the continuous phase, the MSK signal can be coherently demodulated with an improvement in performance over conventional FSK.

As a phase modulation technique, MSK can be thought of as OQPSK with sinusoidal weighting of the symbols.

$$y_k(t) = m_{Ik} \cos\left(\frac{\pi t}{2T}\right)\cos(\omega_0 t) + m_{Qk} \sin\left(\frac{\pi t}{2T}\right)\sin(\omega_0 t) \qquad (5.19)$$

where

m_{Ik} = the in-phase modulation
m_{Qk} = the quadrature phase modulation

The modulator is shown in Figure 5.30 as a direct implementation of a quadrature modulator. Using standard trigonometric identities, the modulation can also be expressed as

$$y_k(t) = \cos\left(\omega_0 t + a_k \frac{\pi t}{2T} + \theta_k\right) \qquad (5.20)$$

where

$a_k = m_{Ik} m_{Qk}$
$\theta_k = 0$ if $m_{Ik} = +1$
$\quad = \pi$ if $m_{Ik} = -1$

In this form, MSK can be seen as a form of FSK with a peak-to-peak deviation ratio of 0.5. The phase trajectory is illustrated in Figure 5.31 for a particular sequence of symbols.

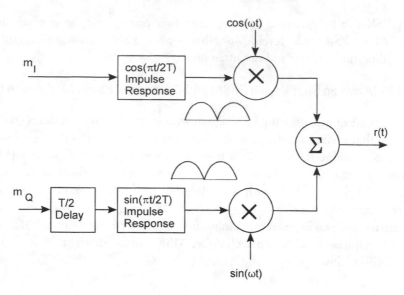

Figure 5.30 MSK modulator.

The power spectrum of MSK is determined by the sinusoidal weighting of the symbols and is

$$P_{MSK}(f) = \frac{16T}{\pi^2}\left(\frac{\cos(2\pi fT)}{16f^2T^2}\right)^2 \tag{5.21}$$

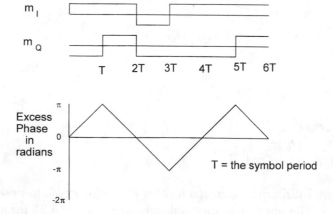

Figure 5.31 MSK phase trajectories.

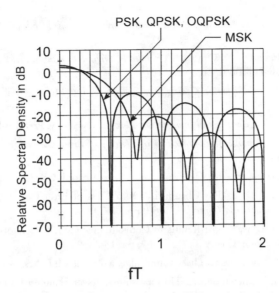

Figure 5.32 Power spectral density of PSK, QPSK, OQPSK, and MSK (shifted to baseband).

The power spectrum, shifted to baseband, of MSK is compared with PSK, QPSK and OQPSK in Figure 5.32. Note, the central lobe is wider than PSK, but the side lobes are much lower.

The concepts behind MSK can be extended to more general types of continuous phase modulation (CPM). To reduce the spectral bandwidth, the MSK symbols have been precoded using duobinary. Just as duobinary signaling can be generalized to classes of partial response signals using precoders (correlative coding), a number of precoding schemes have been investigated [17] using the basic MSK modulation as a starting point. These techniques do suppress out-of-band side lobes at the expense of increased receiver complexity.

References

[1] Nyquist, H., "Certain Topics in Telegraph Transmission Theory," *AIEE Transactions*, 47, 1928.

[2] Siller, Jr., C. A., W. Debus, T. L. Osborne, "Spectral Shaping and Digital Synthesis of an M-ary Time Series," *IEEE Communications Magazine*, February 1989, pp. 15-45.

[3] Lender, A., "The Duobinary Technique for High Speed Data Transmission," *IEEE Transactions on Communication Electronics*, 36, 1963, pp. 61-65.

[4] Waggener, Bill, *Pulse Code Modulation Techniques*, New York: Van Nostrand Reinhold, 1995.

[5]   Peterson, W. W., and E. J. Weldon, *Error-Correcting Codes,* 2nd Ed., Cambridge, MA: The MIT Press, 1984.

[6]   Viterbi, A. J., and J. K. Omura, *Principles of Digital Communication and Coding,* New York: McGraw-Hill Book Company, 1979.

[7]   Berlekamp, E. R., *Algebraic Coding Theory,* New York: McGraw-Hill Book Company, 1968.

[8]   Hamming, R. W., *Coding and Information Theory,* New Jersey: Prentice-Hall, 1980.

[9]   Arazi, B., *A Commonsense Approach to the Theory of Error Correcting Codes,* Cambridge, MA: The MIT Press, 1988.

[10] Pretzel, O., *Error-Correcting Codes and Finite Fields,* Oxford, UK: Clarendon Press, 1992.

[11] Bose, R. C. and D. K. Ray-Chauduri, "On a Class of Error Correcting Binary Group Codes," *Information and Control,* 3, 1960, pp. 68-79.

[12] Lin, S. and H. Lyne, "Some Results in Convolutional Code Generators," IEEE Transactions on Information Theory, IT-13, 1967, pp. 134-139.

[13] Forney, G. D., *Concatenated Codes,* Cambridge, MA: The MIT Press, 1966.

[14] Berrou, C., A. Glavieux, and A. Thitimasjshima, "Near Shannon Limit Error-correcting Coding and Decoding," *ICC '93 Proceedings,* Geneva, Switzerland, May 1993, pp. 1064-1070.

[15] Berrou, C. and A. Glavieux, "Near Optimum Error Correcting and Decoding: Turbo Codes," *IEEE Transactions on Communications,* 44, October 1996, pp. 1261-1271.

[16] Papoulis, A., *The Fourier Integral and Its Applications,* New York: McGraw-Hill Book Co., 1962.

[17] Sundberg, Carl-Erik, "Continuous Phase Modulation," *IEEE Communications Magazine,* 24, April 1986, pp. 25-35.

6

Demodulation and Detection

The performance of a PCM system is determined by the demodulation and detection functions shown in Figure 6.1. The system designer has control over the choice of coding and modulation and can configure the system to meet the performance requirements. There are many applications in which the designer of the receiving equipment has no control over the type of coding and modulation used, and thus must optimize performance within these constraints. In some cases, the constraints may conflict with the desired performance and the designer must make the best of a bad situation.

Although systems using modulated carriers are probably more common than not, baseband signal detection will be discussed first, followed by the discussion of modulated carrier systems. If modulated carriers are coherently demodulated, the modulation/demodulation process is "transparent" and the baseband model can be used for performance analysis. For this case, the transmission channel distortion can be represented by a baseband model. In the case of incoherent demodulation, the performance analysis must directly consider the demodulation process.

This chapter follows the traditional path of first considering the performance in additive, white Gaussian noise (AWGN) followed by the effects of real channel perturbations, such as fading and non-Gaussian interference. There are many excellent texts covering detection and demodulation and no attempt is made to re-run well traveled paths. Communication books, which the author has used for general reference over the past 30 plus years, are listed in [1–16.] Some have gathered dust over the years while others are well worn. Basic con-

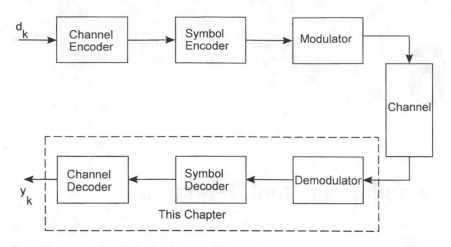

Figure 6.1 Demodulation and detection.

cepts are reviewed and important results summarized, both as equations (where closed form solutions exist) and as charts.

6.1 Detection Basics

Information is transmitted in the PCM system by discrete symbols. At the receiving end, the individual symbols must be detected in the presence of signal distortion and noise. A simple model of the detection system is shown in Figure 6.2.

In a classic paper, Nyquist [26] showed that it was possible to communicate over ideal bandlimited channels at a rate equal to twice the channel cutoff bandwidth. In other words, it is possible to transmit two symbols per second per hertz of bandwidth. An ideal, bandlimited channel has a *sinc(x)* impulse response. If the output of the channel is sampled at a symbol rate of twice the frequency cutoff of the channel, the output has a peak response at the center of the *sinc(x)* impulse response and is zero at all other sampling times. There is no intersymbol interference between the current input symbol and all previous and successive symbols when sampled at the symbol rate. If binary data is transmitted over the Nyquist channel, the communications efficiency would theoretically be 2 bits/sec-Hz.

Although the ideal Nyquist channel is interesting, it is not practical to implement. It is difficult to implement an approximation to the sharp cutoff filter requirements for the Nyquist signal. Furthermore, if there is any timing uncertainty, the basic Nyquist channel has essentially zero timing margin mak-

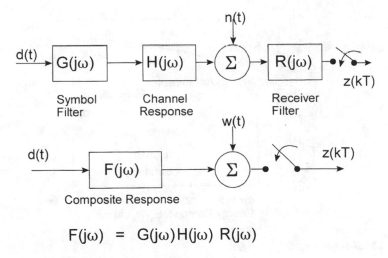

Figure 6.2 Equivalent communication model.

ing it useless for any practical implementation. However, Nyquist also showed that it was not necessary to have a "brick wall" bandlimited channel to satisfy the requirement for zero intersymbol interference. Any filter which exhibits odd symmetry about the half amplitude point also exhibits zero intersymbol interference. This allows filters with excess bandwidth to be used while retaining zero intersymbol interference.

System noise ultimately limits the ability of the receiver to make correct symbol decisions. The Shannon bound [27] establishes a limit to the communications efficiency as a function of signal-to-noise ratio. The Shannon bound relates the communication efficiency expressed in bits/second-Hz to signal-to-noise ratio. It is common practice to measure signal-to-noise ratio as the "energy contrast" ratio, E_b/N_0. If several bits of information are conveyed by one symbol, the symbol energy-to-noise-spectral density is given by:

$$\frac{E_s}{N_0} = \frac{E_b}{N_0} \log_2 M \qquad (6.1)$$

where

M = the number of symbols

The bound, shown in Figure 6.3, assumes signals with infinite time-bandwidth products ($2WT$).

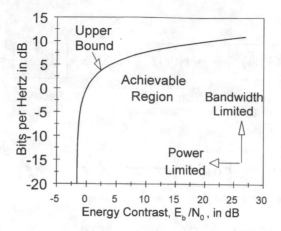

Figure 6.3 Shannon bound.

The symbol energy-to-noise spectral density is equivalent to the signal-to-noise ratio measured in a bandwidth equal to the symbol rate. As an example, to achieve a communication efficiency of 10 bits/sec per Hz requires an energy contrast ratio of 20 dB, or a signal-to-noise ratio of 30 dB in a bandwidth equal to the symbol rate. Thus, if a system has a signal-to-noise ratio of 20 dB with a 100Kbit/sec rate in a bandwidth of about 100 kHz, a 30 dB signal-to-noise ratio in the same bandwidth is required to transmit 1 Mbit/second.

The Shannon bound provides a goal for efficient PCM systems. Efficient signal coding and optimum detection are required to approach the Shannon bound. Viterbi [11] provided a critical component in the search for optimum performance. Viterbi's contribution was an elegant method of decoding convolutional codes using a sequence estimator. The same method later generalized by Forney [28] and others [29] to optimally detect PCM signals with intersymbol inteference. These methods will be discussed in more detail later in this chapter.

Before getting too deeply involved with the detection of PCM signals, a brief review of terminology and detection basics are in order. The simplest PCM detection problem is to decide which of two possible symbols has been transmitted each symbol time. The symbols will be assumed to be statistically independent and the symbol waveform confined to one symbol period. The received signal is assumed to be corrupted by additive, white Gaussian noise as shown in Figure 6.4.

Statistical independence of the symbols and constraining of the symbol to one symbol period implies that knowledge of previous or future symbols cannot improve our decision. White noise is a stochastic process with a constant power

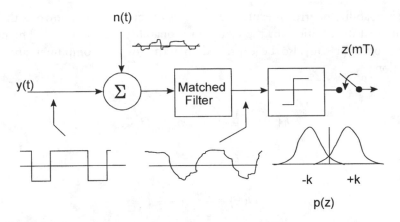

Figure 6.4 Detection model.

spectral density or, equivalently, with a correlation equal to an impulse function. Gaussian noise has a probability density function described by

$$p(x,m,\sigma) = \frac{1}{\sqrt{2\pi}\sigma} e^{\frac{-(x-m)^2}{2\sigma^2}} \tag{6.2}$$

where

m = the mean of the density
σ = the standard deviation

Within these constraints, the optimum detector is a "matched" filter which has an impulse response with a time-reversed replica of the symbol waveform. The matched filter is sampled at the end of the symbol period and the output is compared to a threshold level. If the transmitted symbol has two levels and the sample exceeds the threshold, a "one" is decided. If the sample is less than the threshold, a "zero" is decided. If the symbols are symmetric about zero, the detected samples have values $\pm k$ and the probability of a decision error is

$$P_e = \frac{1}{\sqrt{2\pi}} \int_{\frac{k}{\sigma}}^{\infty} e^{-\frac{u^2}{2\sigma^2}} du = Q\left(\frac{k}{\sigma}\right) \tag{6.3}$$

The probability of error is expressed in terms of the function known as the upper normal distribution, or the Gaussian probability function [30]. The probability of error can also be expressed in terms of the complimentary error function

$$erfc(x) = \frac{1}{\sqrt{\pi}} \int\limits_{x}^{\infty} e^{-u^2} du = 2Q(\sqrt{2}x) \qquad (6.4)$$

$$Q(x) = \frac{1}{2} erfc\left(\frac{x}{\sqrt{2}}\right) \qquad (6.5)$$

For consistency, the $Q(x)$ function will be used in this book. The sampled output of the matched filter is proportional to

$$k = \int s(\tau)h(t-\tau)d\tau = \langle ss* \rangle = E_s \qquad (6.6)$$

where

$\langle s\, s* \rangle$ is the inner product and equals the signal energy, E_s

The root mean square noise, σ, is proportional to the product of the noise spectral density, N_0, and the reciprocal of the symbol period, T.

$$\sigma^2 = \frac{N_0}{2T} \qquad (6.7)$$

Thus the argument of the error function is

$$\frac{k}{\sigma} = \sqrt{\frac{2E_s}{N_0}} \qquad (6.8)$$

where

E_s is the symbol energy

For this case, the probability of error is a function of E_b/N_0, the energy per bit to the noise spectral density. The ratio, E_b/N_0, is also called the energy contrast ratio and will be abbreviated

$$\frac{E_b}{N_0} \equiv \gamma \qquad (6.9)$$

In general, the error performance of all PCM systems will be a function of this ratio and the performance of each system will be compared as a function of this ratio.

Signals can be represented as vectors in an N-dimensional space [4] and a "distance" can be defined between two signals as illustrated in Figure 6.5. Noise is a vector added to the tips of the vectors, and if the noise exceeds one half of the distance between the vectors a decision error will be made. When two signals are opposite one another, the distance is greatest and the performance will be optimum. Two opposite signals are called "antipodal." When represented as vectors, convolution is equivalent to the dot, or inner, product between the vectors. For a PCM system which uses one of two possible symbols to represent ones and zeros, the received signal can be represented by

$$\vec{r} = \vec{s}_i + \vec{n}, i = 0,1 \qquad (6.10)$$

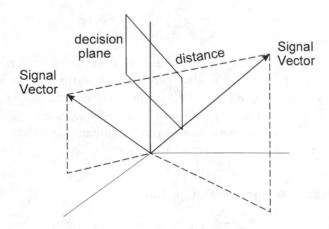

Figure 6.5 Vector representation of signals.

The optimum detector is known to be a "matched" filter, matched to the difference between the two possible signals

$$\vec{h} = (\vec{s}_0 - \vec{s}_1)$$ (6.11)

The output of the matched filter is

$$\vec{z}_i = \langle \vec{h}\vec{r} \rangle = \langle (\vec{s}_0 - \vec{s}_1)(\vec{s}_i + \vec{n}) \rangle$$ (6.12)

For signal, s_0

$$\vec{z}_0 = \langle \vec{s}_0 \vec{s}_0 \rangle - \langle \vec{s}_1 \vec{s}_0 \rangle + \vec{w}$$ (6.13)

For signal, s_1

$$\vec{z}_1 = \langle \vec{s}_0 \vec{s}_1 \rangle - \langle \vec{s}_1 \vec{s}_1 \rangle + \vec{w}$$ (6.14)

where

w is the noise component

The inner product of a vector with itself is the signal energy. Assuming both signals have the same energy, the distance between the two signals is

$$\vec{d} = 2E_s - \langle \vec{s}_1 \vec{s}_0 \rangle - \langle \vec{s}_0 \vec{s}_1 \rangle + noise$$ (6.15)

If the signals are antipodal, $s_0 = -s_1$, so that the distance (without noise) is $4E_s$. If the signals are orthogonal (the inner product is zero), the distance is only $2E_s$. Thus, antipodal signaling requires 3 dB less power than orthogonal signals. The concept of distance will be very important in comparing different PCM signaling methods. The high signal-to-noise ratio performance of different signaling methods will be a direct function of their relative distances.

6.1.1 Baseband Versus Carrier Signaling

Many communication channels are passband with a frequency response band displaced from dc. Baseband signals must be translated to the passband by modulation. In the case of cable channels, the response extends to low frequencies but dc and frequencies near dc are often blocked by transformers or capacitors. Baseband signals with no dc content such as biphase or AMI can be used without the need for modulation. Even in many cable applications, modulation is used to translate baseband signals to specific bands within the cable bandwidth.

A bandpass pulse amplitude modulated (PAM) signal can be represented in terms of a baseband signal (which may be complex) by

$$y(t) = \sqrt{2} \, \text{Re} \left[e^{j\omega_c t} \sum_{-\infty}^{+\infty} d_k f(t - kT) \right] \tag{6.16}$$

where

d_k are the data values
$f(t)$ is the symbol waveform
ω_c is the carrier frequency

The modulator can be represented as shown in Figure 6.6.

The bandpass signal can be AM-DSB, AM-SSB or QAM. The double sideband, amplitude modulation (AM-DSB) is inefficient in the sense that the two sidebands contain redundant information. Single sideband AM (AM-SSB) is twice as efficient (in bits per second-Hz) but is more difficult to generate and demodulate. Quadrature amplitude modulation (QAM) is a commonly used PCM modulation and can be represented by a complex baseband signal, $u(t)$, with the real part representing an in-phase bit stream and the imaginary part representing a quadrature (90 degree phase shift) bit stream as shown in Figure 6.7.

The generalized demodulator for the PAM signal is shown in Figure 6.8. In the case of QAM, the demodulator is implemented as shown in Figure 6.9 with an in-phase and a quadrature phase receiver. If the channel frequency

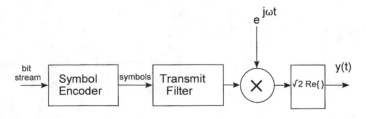

Figure 6.6 Generalized PAM modulator.

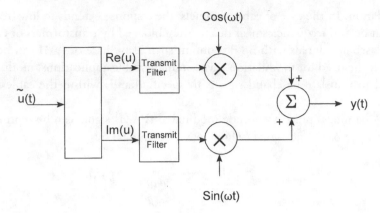

Figure 6.7 QAM modulator.

response is a constant, linear phase within the passband, the baseband symbols can be demodulated without distortion using an ideal frequency reference. In this case, the channel is essentially transparent to the baseband signaling and a pure baseband model can be used for analysis.

For a nonideal channel, the channel can be replaced by an equivalent baseband channel as illustrated in Figure 6.10. The equivalent baseband response of the channel is obtained by linearly translating the channel response to dc. In general, the translated response is not complex conjugate symmetric and the impulse response will be complex. For a complex impulse response, both the in-phase and quadrature bit streams will suffer distortion with crosstalk between the streams.

Phase shift keying (PSK) can be represented by signal points in a two dimensional constellation in a manner similar to QAM. Four phase PSK (QPSK) is equivalent to a four level QAM signal. Demodulation of M-ary PSK can be implemented with a complex demodulator similar to the generalized QAM demodulator.

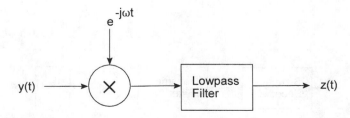

Figure 6.8 PAM demodulator.

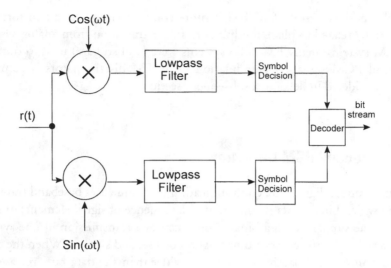

Figure 6.9 QAM demodulator.

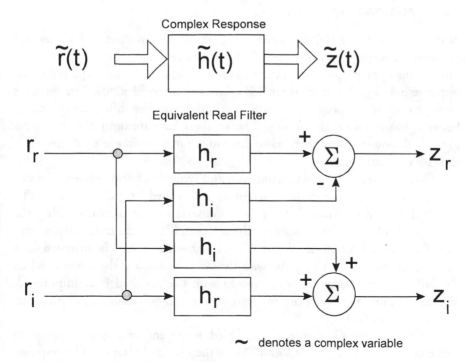

$\sim$ denotes a complex variable

Figure 6.10 Equivalent baseband channel.

Demodulation of PCM/FM is more complex although certain forms of FSK can be related to phase modulation. In the transition from analog systems to PCM systems, many FM radio systems were (and still are) used to transmit baseband PCM signals. Demodulation using FM discriminators is common and is considered in detail in a subsequent section.

6.2 Baseband PCM Detection

The most direct digital data communication technique uses baseband transmission in which binary data is converted to a sequence of signal elements or symbols. In the simplest case, a binary "one" causes a transmission of one symbol while a binary "zero" causes a transmission of a second symbol. When the communication channel bandwidth is much wider than the data rate, rectangular symbols can be used for data transmission without distortion. A number of baseband symbols were introduced in Chapter 5.

6.2.1 Rectangular Symbols

Nonreturn-to-zero (NRZ) signaling is the simplest of all these techniques and represents a binary "one" by a pulse of one polarity and a binary "zero" by a pulse of the opposite polarity. If the two symbols have the same shape and are of opposite polarity, the signal is termed a binary, antipodal signal. The bit error performance in the presence of noise depends upon the difference in energy between the transmitted symbols. The symbols have maximum energy for a given peak power if they are rectangular and the maximum energy difference if they have opposite polarity.

With nonbandlimited transmission and symbols of finite duration, it can be shown that the optimum detector is a so-called "matched filter." The matched filter both maximizes signal-to-noise ratio and minimizes bit error probability. The impulse response of the matched filter is the time-reversed replica of the symbol waveform. For an NRZ square-sided pulse, the matched filter impulse response is a square pulse with the same duration as the symbol. When the NRZ symbol waveform is convolved with the matched filter impulse response, the output of the optimum detector is a triangular shaped pulse as illustrated in Figure 6.11.

The matched filter output is sampled at the end of a symbol period to make the optimum symbol decision. Since the matched filter impulse response is a rectangular pulse, the frequency spectrum of the detector has a *sinc(x)* fre-

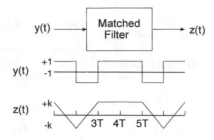

Figure 6.11 NRZ matched filter.

quency response with the first null frequency at the symbol rate. If the gain of the matched filter is chosen such that the matched filter output for an NRZ signal has peak values of $\pm k$ at the optimum sampling points, the probability of a symbol error is given by

$$P_{NRZ-L} = Q\left(\frac{k}{\sigma}\right) \tag{6.17}$$

Expressing the filter output in terms of the symbol energy, the error probability can alternatively be expressed in terms of the energy contrast ratio as

$$P_{NRZ-L} = Q\left(\sqrt{\frac{2E_b}{N_0}}\right) \tag{6.18}$$

In the case of the NRZ PCM with binary data, the symbols take on only two values and convey one bit of information per symbol. Therefore, the symbol error probability is equal to the bit error probability.

The same method can be applied to the analysis of the other rectangular waveform symbols. The bit error probability of many of the common rectangular symbols is summarized in Table 6.1.

The bit error probability is plotted in Figures 6.12 and 6.13 for binary valued rectangular symbols. Additional levels can be used with NRZ symbols to convey multiple bits per symbol. This is called pulse amplitude modulation (PAM) and the bit error probability for 4-level (2 bits per symbol) and 8-level (3 bits per symbol) is shown in Figure 6.14.

Table 6.1
Bit Error Probability of Common Rectangular Symbols

Signaling	Bit Error Probability	Remarks/reference
NRZ-L, Biphase-L	$Q\left(\sqrt{2\gamma}\right)$	
NRZ-M, NRZ-S, NRZ-I, Biphase-M,S	$2Q\left(\sqrt{2\gamma}\right)$	approximate
AMI (100% pulse width)	$Q\left(\sqrt{\gamma}\right)$	approximate
AMI (50% pulse width)	$Q\left(\sqrt{\dfrac{\gamma}{2}}\right)$	approximate
Delay Modulation (DM)	$Q\left(\sqrt{\gamma}\right)$	single bit decision [16]
DM	$\dfrac{1}{2}Q\left(\sqrt{\gamma}\right)+\dfrac{1}{3}Q\left(\sqrt{2\gamma}\right)+\dfrac{1}{6}Q\left(\sqrt{3\gamma}\right)$	two bit decision [16]
4-level PAM	$1.5Q\left(\sqrt{\dfrac{4\gamma}{5}}\right)$	reference [8]
8-level	$Q\left(\sqrt{\dfrac{18\gamma}{63}}\right)$	approximate (high SNR) [15]

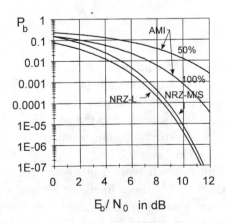

Figure 6.12 NRZ and AMI bit error probability.

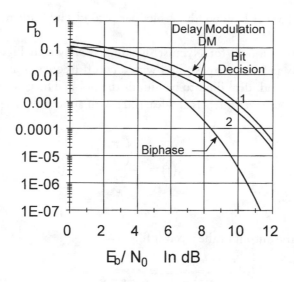

Figure 6.13 Biphase and DM bit error probability.

6.2.2 Nyquist Symbols

If the channel is bandlimited, rectangular symbols will be distorted and intersymbol interference introduced. As indicated previously, an ideal bandlimited channel (a Nyquist channel) has a *sinc(x)* impulse response with zeros at multiples of one half the cutoff frequency. It is theoretically possible to signal

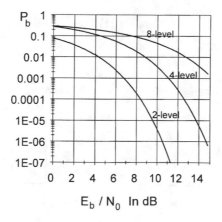

Figure 6.14 PAM bit error probability.

at twice the channel bandwidth (2 bits per second-Hz) with no intersymbol interference.

A detection model for Nyquist signaling is shown in Figure 6.15. The symbol filter is an ideal lowpass filter with a cutoff frequency at $1/2T$. The receiver uses a second ideal filter to bandlimit the noise. The cascade response of the symbol and receiver filters is an ideal filter with an impulse response

$$h_{NYQ}(t) = \frac{\sin\left(\frac{\pi t}{T}\right)}{\frac{\pi t}{T}} \tag{6.19}$$

The signal at the output of the receiver filter is

$$r(t) = \sum_{-\infty}^{+\infty} s_k h(t - kT) \tag{6.20}$$

where

$$s_k = \pm 1$$

The mean square noise at the output of the receiver filter is

$$\sigma^2 = N_0 \frac{1}{2T} \tag{6.21}$$

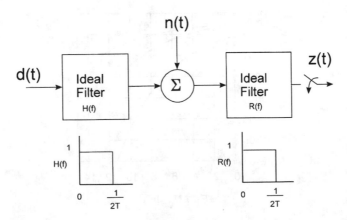

Figure 6.15 Nyquist signal detection model.

The sample signal-to-noise ratio is

$$SNR = \frac{2s_k^2 T}{N_0} = \frac{2E_b}{N_0} \qquad (6.22)$$

From the signal-to-noise ratio, the bit error probability is exactly equal to that of NRZ detected using a matched filter.

$$P_{NYQ} = Q\left(\sqrt{2\gamma}\right) \qquad (6.23)$$

Although the ideal Nyquist channel is not realizable, it suggests a different approach to signaling over bandlimited channels. A filter exhibiting odd symmetry about the half amplitude point also exhibits zero intersymbol interference. By accepting some excess bandwidth over the Nyquist rate, practical realizations of Nyquist signaling can be achieved. The raised cosine, as discussed in Chapter 5, is one of the most commonly used Nyquist signals.

Two detection models can be used for the raised cosine signal, as shown in Figure 6.16. The first model is similar to the ideal Nyquist signal detector, using a symbol filter to form the raised cosine symbols and an ideal bandlimited filter for the detection filter. The second model is more realistic, splitting the raised cosine response between the transmitter and the receiver. The transmitter and receiver use a filter response, $H^{\frac{1}{2}}(f)$, which is the square root of the overall desired response. In both models, the total channel response satisfies the Nyquist criteria and, even though the noise spectral density at

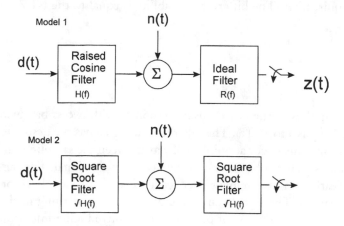

Figure 6.16 Raised cosine detection model.

the output of the second model is nonwhite, the noise samples are statistically independent.

The overall impulse response of the system is

$$h_{RC}(t) = \frac{\sin\dfrac{2\pi t}{T}}{\dfrac{2\pi t}{T}\left(1-\left(\dfrac{2t}{T}\right)^2\right)} \qquad (6.24)$$

The peak signal at the sampling time is

$$r(0) = s\,h(0) = s \qquad (6.25)$$

The mean square noise at the output of the square root detection filter is computed from the autocorrelation function.

$$R(\tau) = \frac{N_0}{2}\int\limits_{-\infty}^{+\infty}\left|H^{1/2}(f)\right|^2 e^{j2\pi f\tau}\,df$$

$$R(\tau) = \frac{N_0}{2}\int H(f)e^{j2\pi f\tau}\,df = \frac{N_0}{2}h(\tau) \qquad (6.26)$$

The mean square noise is $R(0)$ and the sampled signal-to-noise ratio is identical to the Nyquist filter. The bit error probability is equal to the NRZ error probability.

$$P_{RC} = Q\!\left(\sqrt{2\gamma}\right) \qquad (6.27)$$

This result applies to the raised cosine responses with excess bandwidths from 0% (ideal Nyquist) to 100%. The raised cosine signal has the desirable property that transitions only occur at points half way between the sampling times, therefore a timing signal can be accurately derived from the signal. The raised cosine signal is clearly an improvement over rectangular shaped symbols for the bandlimited channel. The signal can be bandlimited to something in the order of the symbol rate without loss of performance or introducing intersymbol interference between symbols. The signal exhibits excess bandwidth over the

Nyquist criteria and cannot fully achieve the two symbols per second per Hz efficiency of an ideal Nyquist signal.

6.2.3 Partial Response Signals

The partial response signals, presented in Chapter 5, introduce a controlled amount of intersymbol interference to shape the signal spectrum and control its bandwidth. Since the intersymbol interference was introduced deliberately, it is known at the receiver and its effects can be canceled. The duobinary signal is used to introduce the bit error performance of partial response signaling. A detection model for duobinary is shown in Figure 6.17.

The impulse response of this channel is readily shown to be of the following form

$$h(t) = \frac{\pi}{4}\left[\frac{\sin\pi\left(\frac{t}{T}+\frac{1}{2}\right)}{\pi\left(\frac{t}{T}+\frac{1}{2}\right)} + \frac{\sin\pi\left(\frac{t}{T}-\frac{1}{2}\right)}{\pi\left(\frac{t}{T}-\frac{1}{2}\right)}\right] \tag{6.28}$$

The impulse response consists of the sum of two *sinc(x)* functions and is nonzero at two points.

$$h\left(\pm\frac{T}{2}\right) = \frac{\pi}{4}$$

$$h\left(\pm\frac{kT}{2}\right) = 0 \quad k \neq 1 \tag{6.29}$$

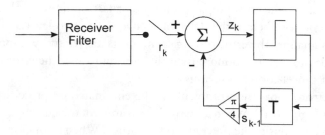

Figure 6.17 Duobinary detection model.

A controlled amount of intersymbol interference is introduced between symbols with the intersymbol interference component having the same amplitude as the direct component. By deliberately introducing this intersymbol interference, the bandwidth of the baseband signal is limited to the Nyquist bandwidth making it possible to transmit two symbols/sec-Hz. The penalty for achieving this signaling efficiency is a degradation in signal-to-noise ratio due to the deliberate intersymbol interference. Since the intersymbol interference is known, a decision directed cancellation of the interference component can be achieved at the receiver. If the baseband channel is sampled at the symbol rate, the detection filter output is

$$r_k = \frac{\pi}{4} s_k + \frac{\pi}{4} s_{k-1} + n_k \tag{6.30}$$

Suppose that the previous symbol received, s_{k-1}, has been correctly detected and is known to the receiver. This signal can then be subtracted from the received signal

$$z_k = r_k - \frac{\pi}{4} s_{k-1} = \frac{\pi}{4} s_k + n_k \tag{6.31}$$

Symbol decisions can be made based on this level. A decision threshold is placed at zero and a "one" is decided if the decision statistic is positive and a "zero" if the statistic is negative. At high signal-to-noise ratios, the bit error probability for duobinary is

$$P_{duo} = Q\left(\sqrt{\frac{\pi^2 \gamma}{8}}\right) \tag{6.32}$$

If an incorrect bit decision is made, the decision for the next symbol will be in error. Fortunately, the error propagation is not catastrophic and only short error bursts are produced before the detector recovers. For two-level and four-level duobinary signaling, the error propagation effects are relatively minor with the average error probability no more than three or four times the error probability with correct cancellation of previous elements.

The error propagation problem can be eliminated by precoding the data as shown in Figure 6.18. When a "zero" is transmitted, the current coder output is chosen to be opposite the previous coder output. When a "one" is to be transmitted, the coder output is equal to the previous coder output.

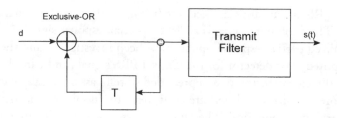

Figure 6.18 Precoded duobinary.

When a binary zero is transmitted, the received output is at level zero. When a binary one is transmitted, the received level is either $\pm \pi k/2$. The receiver decides a binary one if the received symbol is greater than $+\pi\ k/4$ or less than $-\pi k/4$. Otherwise, a binary zero is decided. The relative decision level is the same as the decision-directed detector. In this case, however, decisions are made independently from symbol to symbol and no error propagation effects can occur.

As discussed previously, the duobinary technique can be generalized to types of signals called partial response signaling (PRS). Recognizing that the duobinary symbol is the weighted sum of two *sinc(x)* functions delayed by one symbol period, an obvious extension of the technique creates signals which are weighted sums of *sinc(x)* signals separated by multiples of the symbol period. The generalized PRS uses a finite impulse response (FIR) digital filter followed by an ideal Nyquist filter as shown in Figure 6.19. The ideal Nyquist filter produces the *sinc(x)* symbol waveforms and the FIR filter provides the weighted combination of these waveforms.

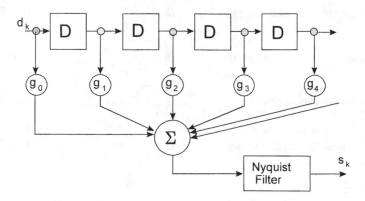

Figure 6.19 Generalized partial response signal.

The PRS signals are classified according to the transfer function of the FIR filter. The duobinary signal is a Class 1 partial response signal. A number of the generalized partial response signals have been investigated and the performance computed. The detector for the Class 4 PRS signal can be implemented as a decision directed detector, or as a precoded system, as in the case of duobinary. Furthermore, the decision levels are equivalent to duobinary and therefore the error performance is identical. The bit error performance of the Nyquist and PR signals is summarized in Table 6.2 and shown in Figure 6.20.

The use of precoding is not limited to binary signals. The partial response concept can be extended even further by replacing the ideal Nyquist filter with a filter satisfying the generalized Nyquist criteria with odd symmetry around the cutoff frequency.

Achieving the theoretical bit error performance presupposes accurate sampling. In reality, timing variations occur due to noise, signal distortion, and other factors. Some baseband signals offer better timing margins than others. The timing margin is best illustrated by considering the "eye" pattern for a given signal. The eye pattern is obtained by superimposing segments of the received waveform on a common time reference. Typically, a segment of two to four symbol periods is used as a reference. The eye pattern for the matched filter

Table 6.2
Bit Error Performance of Nyquist and PR Symbols

Signaling	Bit Error Probability	Remarks
Raised Cosine, Binary	$Q\left(\sqrt{2\gamma}\right)$	
Raised Cosine, Quaternary	$Q\left(\sqrt{\dfrac{4\gamma}{5}}\right)$	Approximate
Partial Response Class 1,4 Binary	$Q\left(\sqrt{\dfrac{\pi^2\gamma}{8}}\right)$	High SNR
Partial Response Class 1,4 Quaternary	$Q\left(\sqrt{\dfrac{\pi^2\gamma}{20}}\right)$	High SNR
Partial Response Class 2, Binary	$Q\left(\sqrt{\dfrac{\gamma}{2}}\right)$	High SNR
Partial Response Class 2, Quaternary	$Q\left(\sqrt{\dfrac{\gamma}{5}}\right)$	High SNR
where $\gamma = \dfrac{E_b}{N_0}$		

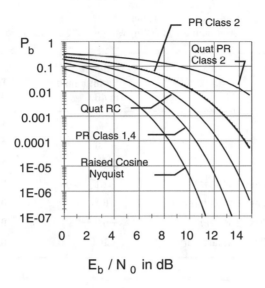

Figure 6.20 Bit error probability for Nyquist and partial response signaling.

output of an NRZ signal is shown in Figure 6.21 over a four symbol period. All possible symbol combinations form the eye pattern. The ideal sampling time and decision level is at the center of the eye.

The relative bit error performance and timing margins are summarized in Table 6.3 for a number of the baseband signals.

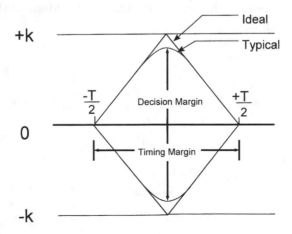

Figure 6.21 Eye pattern for NRZ signal.

Table 6.3
Bit Error Performance and Timing Margins of Baseband Signals

Signaling	Relative Performance in dB	Timing Margin (Relative to $\pm$ $T/2$)
NRZ	0	1.0
4-level PAM	-4.0	0.33
Raised Cosine	0	1.0
Partial Response, Class 1,4 Binary	-2.1	0.67
Partial Response, Class 1,4 Quaternary	-6.1	0.18
Partial Response, Class 2, Binary	-6.0	0.36
Partial Response, Class 2, Quaternary	-10.0	0.18

6.3 Coherent Demodulation

Amplitude modulation is equivalent to multiplication by a complex exponential carrier. Demodulation is accomplished by multiplying the received signal by a reference complex carrier and removing higher order frequency components as shown in Figure 6.22. When the reference carrier has the same frequency and phase as the modulating carrier, the demodulation is said to be coherent. Obtaining a coherent reference carrier can be achieved by either transmitting an unmodulated portion of the carrier or by extracting an estimate of the carrier from the received signal. Carrier synchronization is considered in the next chapter. In this section, it is assumed that a coherent reference is available.

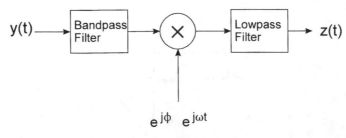

Figure 6.22 Coherent demodulation.

6.3.1   Single Sideband Amplitude Modulation (SSB/AM)

A direct method of implementing a bandwidth efficient modulated carrier system uses single sideband (SSB) modulation. The spectral occupancy of the signal for ideal SSB modulation is identical to that of the baseband signal and the SSB carrier frequency can be chosen to place the signal at some arbitrary point in the frequency band. If ideal SSB demodulation is used, the performance is identical to the performance of the equivalent baseband signal. This would appear to be the ideal modulated carrier technique since the carriers can be chosen to fit the signal at any desired location. In practice, however, the technique is not so straightforward. The rejection of image frequencies and the generation/demodulation of the single sideband carrier can be a significant problem. The demodulation of SSB can be simplified by adding a reference "pilot tone" to the transmitted signal. The pilot tone can be used to recover the baseband signal at the expense of added transmitter power. The performance of an ideal SSB/AM system is equivalent to the performance of the baseband signals used.

6.3.2   Quadrature Amplitude Modulation (QAM)

The quadrature modulation technique employed by QPSK can be extended to a more general quadrature amplitude modulation (QAM). The QAM technique uses in-phase and quadrature phase carriers which are modulated by two baseband signals. The data is recovered by a quadrature demodulator as illustrated in Figure 6.23. By using Nyquist signals on the quadrature AM carriers, the

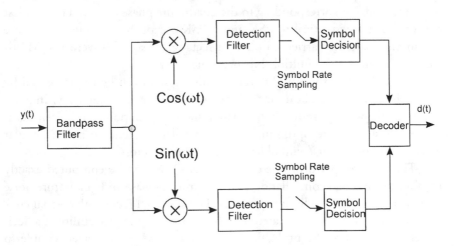

Figure 6.23   QAM demodulator.

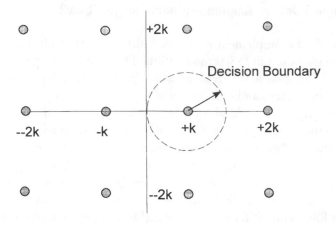

Figure 6.24 Symbol decision in M-ary QAM.

Nyquist signaling rate can be obtained for both the in-phase (I) and the quadrature phase (Q) channels. If a bipolar Nyquist signal amplitude modulates a carrier, a double sideband, suppressed carrier signal is formed with a nominal bandwidth equal to the symbol rate. Modulating a second carrier in phase quadrature with the first carrier and summing the two signals, a signal is formed which still has a bandwidth equal to the symbol rate but signals at the Nyquist rate.

This concept can be extended by transmitting multiple bits per symbol using multilevel signal waveforms. As discussed previously, the QAM signals are represented in a signal space with one axis corresponding to the in-phase carrier and the second axis corresponding to the quadrature phase carrier. The simplest cases of QAM are identical to PSK and QPSK except that Nyquist signals are used to modulate the carriers in place of rectangular signal waveforms. Multilevel signals can convey multiple bits per dimension.

A signal design for a multilevel signaling system can be implemented by placing the signal points at uniform intervals on the rectangular array. In practice, there are more optimum ways of assigning the signal points. For even numbers of bits/symbol the optimum signal constellations tend to be rectangular with odd number of bits/symbol having the form of a cross.

The bit error performance of multilevel QAM can be computed exactly [15] for a given signal constellation assuming the in-phase and quadrature noise components are statistically independent. For a general rectangular signal constellation with the points separated by a distance of $2k$, the probability of a decision error is equal to the probability that noise causes the received sample to exceed a circle with radius, k, about the signal point as illustrated in Figure 6.24.

If the in-phase and quadrature components of the noise are Gaussian random variables, the probability that the radius, k, is exceeded is [31]

$$P_{sym} = e^{-\frac{k^2}{2\sigma^2}} \tag{6.33}$$

The signal energy of a square M-level ($\log_2$ M bits per symbol) constellation is

$$s^2 = \frac{2(M-1)k^2}{3} \tag{6.34}$$

To compare the bit error performance of the QAM signal with other modulated signals, the signal energy must be normalized to the binary case and the energy per bit computed from the energy per symbol.

$$k^2 = \frac{3\log_2 M}{2(M-1)} E_b \tag{6.35}$$

The symbol error probability for a large square signal constellation is approximately

$$P_{sym} = Ce^{\frac{3\log_2 M}{2(M-1)}\gamma} \tag{6.36}$$

where

C = a constant between 1.5 and 4

The bit error probability is approximately the symbol error probability divided by the number of bits per symbol.

$$P_b \approx \frac{P_{sym}}{\log_2 M} = \frac{C}{\log_2 M} e^{-\frac{3\log_2 M}{2(M-1)}\gamma} \tag{6.37}$$

At high signal-to-noise ratios, the multiplicative constant can be assumed to be one without significant error. The performance has been computed for a number of QAM signal constellations and the bit error probability for several QAM signals is summarized in Table 6.4 and plotted in Figure 6.25

The in-phase and quadrature phase carriers can be modulated by any of the baseband signals previously discussed. In order to achieve optimum com-

Table 6.4
Bit Error Probability for QAM Signals

Signal	Bit Error Probability	Remarks
4-level QAM (QPSK)	$Q\left(\sqrt{2\gamma}\right) - 0.5Q^2\left(\sqrt{2\gamma}\right)$	[15]
16-level QAM	$\dfrac{3}{4}Q\left(\sqrt{\dfrac{9}{7}\gamma}\right) - \dfrac{2.25}{4}Q^2\left(\sqrt{\dfrac{9}{7}\gamma}\right)$	[15]
M-ary QAM	$e^{-\frac{3\log_2 M}{2(M-1)}\gamma}$	Approximate, high SNR, large constellation
where $\gamma = \dfrac{E_b}{N_0}$		

munications efficiency, a spectrally efficient modulation should be used. This suggests that partial response signals are ideal for use in the QAM case as well as the baseband case. Since the data is modulated on a carrier, the necessity to limit low frequency components is eliminated and Class 1 duobinary signaling provides a very compact signaling spectrum with a maximum RF bandwidth occupancy of one Hz/symbol/second. If a 16-point symbol constellation (4 bits/symbol) is used, an overall spectral efficiency of 4 bits/second-Hz of bandwidth can be obtained using duobinary modulation.

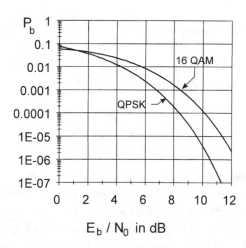

Figure 6.25 QAM bit error probability.

6.3.3 PSK and M-ary PSK

Binary PSK is a special form of amplitude modulation

$$y(t) = \cos(2\pi f_c t) \sum_{-\infty}^{+\infty} d_k p(t - kT) \tag{6.38}$$

where

$d_k = \pm 1$
$p(t) =$ the symbol pulse shape

Coherently demodulating PSK produces the baseband signal

$$r(t) = \sum_{-\infty}^{+\infty} d_k p(t - kT) \tag{6.39}$$

Thus the performance of PSK is identical to bipolar NRZ-L signaling and is an optimal signaling format. Quadrature PSK (QPSK) is the same as 4-level QAM using NRZ-L baseband symbols. The in-phase and quadrature carriers are each binary PSK signals so the bit error probability is the same as binary PSK for the same energy per bit. The energy per symbol is twice that of PSK but the bandwidth is one half that of PSK for the same information rate.

When large number of phases are used, the distance between signal points is reduced as shown in Figure 6.26. The distance relative to the distance between PSK and QPSK symbols is

$$2\sin\frac{\pi}{M} \tag{6.40}$$

where

$M =$ the number of levels

For more than eight levels, the distance is halved for each doubling of the number of levels and the noise margin reduced 6 dB. A tight upper bound for the bit error performance of M-ary PSK is given by

$$P_{MPSK} \leq 2Q\left(\sqrt{2\sin\frac{\pi}{M}\frac{E_s}{N_0}}\right) \tag{6.41}$$

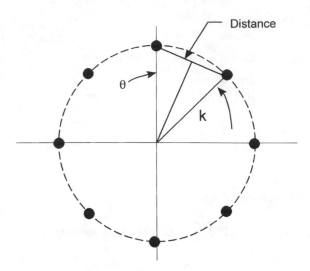

Figure 6.26 M-ary PSK signal distance.

The M-ary PSK is relatively inefficient compared to the Shannon bound for more than four levels.

6.3.4 Frequency Shift Keying (FSK) and Continuous Phase Modulation (CPM)

In binary FSK, a symbol is represented by a tone at one of two frequencies, f_1, or f_2.

$$s(t) = \sqrt{2}A\cos(2\pi f_i t) \qquad i = 0,1 \tag{6.42}$$

If the tone frequencies are chosen to be

$$f_0 = \frac{\dfrac{m}{2}}{T}$$

$$f_1 = \frac{\dfrac{m}{2}+1}{T} \tag{6.43}$$

where

m = an integer

Since one frequency is exactly one cycle higher (or lower) than the second frequency, the phase is continuous between symbols and the peak-to-peak deviation is equal to the symbol rate. The optimum receiver for coherent FSK is shown in Figure 6.27. The peak-to-peak frequency deviation is equal to the bit rate $(1/T)$ and the tones are orthogonal. Because of the orthogonality there is no cross-talk between the detector outputs and the noise is statistically independent. The analysis of the bit error probability is tedious, but direct, and the error probability is

$$P_{CFSK} = Q(\sqrt{\gamma})$$ (6.44)

The bit error probability of coherent is 3 dB poorer than PSK because the FSK tones are orthogonal but not antipodal. The performance of coherent FSK can be improved using an ideal matched filter detector. The matched filter detector is matched to a signal equal to the difference between the two tones. The distance between the two tones can be increased over the orthogonal case using a modulation index of 0.71. In this case, the bit error probability is

$$P_{MF} = Q(\sqrt{1.2\gamma})$$ (6.45)

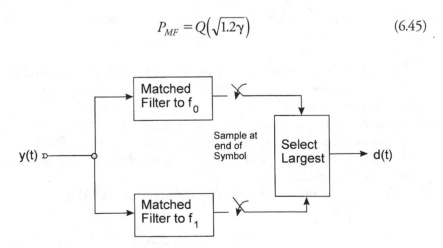

Figure 6.27 Optimum coherent FSK receiver.

Coherent FSK can be extended to M-ary FSK using a receiver with a bank of correlators, one for each frequency. In the case of quaternary (four level) FSK, the bit error probability is approximately

$$P_{QFSK} = Q\left(\sqrt{2\gamma}\right) \qquad (6.46)$$

The error probability is the same as PSK but the QFSK requires about twice the bandwidth of PSK. The M-ary receiver can be extended to even more levels with the error probability remaining approximately the same but the overall bandwidth increasing with the number of levels. At the same time, the information rate is increased by one bit per second-Hz with each doubling of the number of levels.

6.3.4.1 Minimum Shift Keying (MSK)

Minimum shift keying (MSK) [32,33] was originally called "fast FSK" because of its interpretation as continuous phase FSK with a peak-to-peak deviation equal to one half the bit rate. The MSK signal can be represented as

$$y(t) = \sqrt{2}A\cos\left(2\pi f_c t + a_k \frac{\pi t}{2T}\right) \qquad (6.47)$$

This can be expanded as

$$y(t) = \sqrt{2}A\left(\cos\frac{a_k \pi t}{2T}\cos 2\pi f_c t - \sin\frac{a_k \pi t}{2T}\sin 2\pi f_c t\right) \qquad (6.48)$$

In this form, MSK is recognized as staggered QPSK with the data modulating in-phase and quadrature carriers with sinusoidal pulses having a duration of twice the symbol period. The quadrature carrier representation of MSK suggests a coherent demodulator as shown in Figure 6.28. Matched filters at the output of the mixers reject twice frequency components and optimally filter the baseband signals. The two matched filters are sampled every other bit period, staggered by one bit period.

The output of the matched filter for the in-phase channel is

$$z_k = \frac{\sqrt{2}A}{2}\int_0^{2T} s_1(\tau)h(t-\tau)d\tau \qquad (6.49)$$

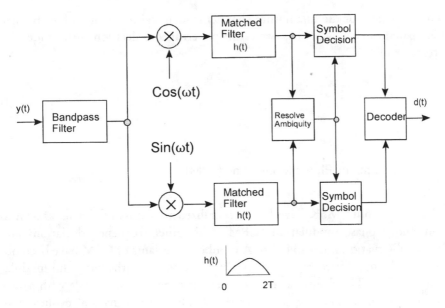

Figure 6.28 Coherent MSK demodulator.

where

$$h(\tau) = \sin \frac{\pi\tau}{2T}, \, 0 \le \tau \le 2T$$
$$= 0, \text{ elsewhere}$$
$$s_I(\tau) = \text{the in-phase signal}$$

The peak signal at the matched filter output is

$$z_k(2T) = \frac{\sqrt{2}}{2} AT \tag{6.50}$$

The mean square noise at the matched filter output is

$$\sigma^2 = \frac{N_0 T}{4} \tag{6.51}$$

The bit error probability for the in-phase channel is

$$P_{IP} = Q\left(\sqrt{2\gamma}\right) \tag{6.52}$$

The error rate for the quadrature channel is identical to the in-phase channel. Compared to PSK, MSK has two errors per erroneous decision with an ideal bit error rate of

$$P_{MSK} = 2Q\left(\sqrt{2\gamma}\right) \tag{6.53}$$

6.3.4.2 Continuous Phase Modulation (CPM)

The ideas behind MSK have been generalized to a class of signals known as continuous phase modulation (CPM) with other frequency deviations and longer observation times [34–39]. A number of variants of CPM have been obtained by using different pulse shapes and durations for the baseband modulation symbols. The thrust of these efforts has been to find signals with a very compact spectra. The use of duobinary symbols was a natural evolution of MSK. As expected, duobinary MSK (DBMSK) reduces the spectral sidelobes even further than conventional MSK. With precoding and conventional detection, DBMSK requires about 2.1 dB more power than MSK in return for the improvement in bandwidth. More advanced maximum likelihood sequence detectors (discussed in a later section) can reduce the performance penalty to about 0.3 dB with an increase in detector complexity.

Generalizing duobinary to correlative coding has resulted in "tamed frequency modulation" (TFM) and generalized TFM [35]. A series of raised cosine time pulses (not to be confused with raised cosine spectrum) have also been proposed in place of the sinusoidal symbol pulses. These variants are called L-RC MSK where L refers to the duration of the symbol, for instance, 2-LC MSK uses a 2 symbol period raised cosine pulse waveform. The variations of CPM can be compared on the basis of spectral efficiency (bits/Hz-second) and relative bit error performance. The spectral efficiency is based on the bandwidth required to pass 99% of the modulated signal power. As an example, PSK requires a bandwidth of 8 times the symbol rate ($BT = 8$) to pass 99% of the signal energy. Since PSK conveys 1 bit per symbol, the spectral efficiency is 0.125 bits/Hz-second.

The bit error performance is compared on the basis of the minimum distance of the signaling method. As a reference, the minimum distance of PSK is two. The minimum distance of the CPM signals can be computed (with great effort) and provide a basis for comparing the signals. The CPM signals are compared in Table 6.5. The bit error performance of the CPM signals are shown in Figure 6.29.

Table 6.5
CPM Signal Comparison

Signaling	Minimum Free Distance	BT	R/B bits/Hz-seconds	Relative Performance in dB
PSK	2	8	0.125	0
MSK	2	1.2	0.83	0
DBMSK	1.724	0.92	1.09	−0.64
TFM	1.453	0.80	1.25	−1.38
2-RC	1.97	1.10	0.91	~0

6.3.5 Spread Spectrum

The simplest (and probably the most common) spread spectrum signal uses coherent PSK or QPSK modulation with a binary pseudorandom chip sequence. The coherent demodulator for this method is shown in Figure 6.30. If the carrier and chip sequence references are ideal, the performance in AWGN is identical to conventional PSK. Spread spectrum systems are designed to reject jamming and nonwhite interference and the DS/PSK system is quite robust in rejecting several types of jamming [40]. The detailed performance of spread spectrum systems is left to texts dealing exclusively with these systems [40–42].

6.3.6 Coherent Demodulation of Optical Signals

Optical transmission is used for most very high bit rate PCM systems. These systems have been called lightwave systems. A laser produces a very high fre-

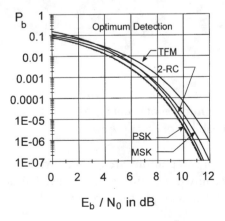

Figure 6.29 Bit error performance of CPM.

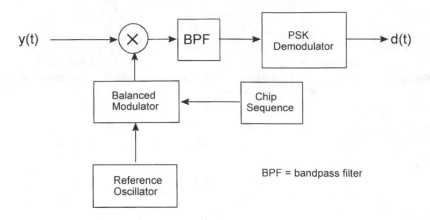

Figure 6.30 DS/PSK demodulator.

quency wave whose frequency and phase stability vary according to the type of lasing device. Lasers can be modulated using PSK, FSK, and on-off keying (OOK); however, the frequency stability limits the types of modulations which are practical.

OOK is the simplest modulation for optical carriers. An optical pulse is transmitted for "one" with no transmission for a "zero." The quantum nature of light places a fundamental limit to the performance. An ideal quantum detector, shown in Figure 6.31, would count photons in each bit. In OOK, a "one" is represented by an optical pulse with no pulse for a "zero." The ideal detector would set a threshold at one photon and any photons in a bit greater than or equal to one would be counted as a "one." The probability density of the detector photocurrent is Poisson

$$p(k) = \frac{m^k e^{-m}}{k!} \tag{6.54}$$

where

m = the mean number of events
$p(k)$ = the probability of exactly k events

With the ideal detector, no photons are received when a "zero" is transmitted. When a "one" is transmitted, N_1 photons are received. A detection error is made if a "one" is transmitted and no photons are detected

$$P_e = p(0) = e^{-N_1} \tag{6.55}$$

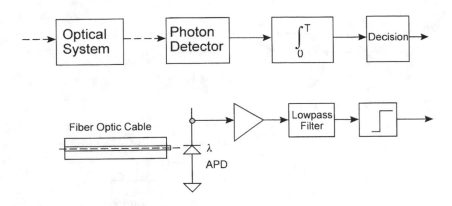

Figure 6.31 Direct optical detector for OOK.

This is a fundamental bound on performance. The ideal detector assumes a quantum efficiency of 1 (every photon produces an electron). Optical communication links are typically judged on the basis of a bit error probability of 10^{-9}. To achieve an error probability of 10^{-9} requires about 20 photons per pulse, or an average optical flux of 10 photons per bit (since a "zero" transmits no power.)

Phase or frequency modulation can be used if the laser has sufficient frequency stability. The frequency stability of lasers is described by the device linewidth. Typical semiconductor lasers exhibit linewidths in the order of 5 to 50 MHz. There are three general techniques for demodulating the lightwave signal depending on the modulation. For PSK modulation, either homodyne or heterodyne detection is used as illustrated in Figure 6.32 [43–45]. In homodyne detection, a phase locked local laser is summed with the signal in an optical combiner and the resulting signal detected using a photodiode. The electric field at the output of the optical combiner is

$$\vec{E}_r = E_s d_k e^{j\omega_c t} + E_{LO} e^{-j\omega_c t}$$

$$d_k = \pm 1 \tag{6.56}$$

where

E_s is the amplitude of the received PSK signal
E_{LO} is the amplitude of the local oscillator (assumed to be phase locked)
d_k is the PSK modulation

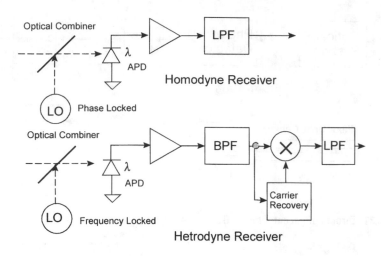

Figure 6.32 Coherent optical receivers.

The photodiode detects the power of the received signal

$$P_r = C\left|\vec{E}_r\right|^2 = P_{DC} + 2E_s E_{LO} d_k \qquad (6.57)$$

The photodiode current contains a dc component proportional to the sum of the signal and local oscillator powers and a component proportional to the baseband modulation. The amplitude of the baseband component is proportional to the level of the local oscillator and can be adjusted so the the shot noise in the detected current is greater than the preamplifier noise.

In order to phase lock the local reference, a small amount of power must be extracted from the signal to be used in an optical phase locked loop. This is quite difficult to achieve and requires linewidth stabilities of 0.05% or less of the bit rate. Despite the difficulty of implementing a homodyne receiver, the bit error performance can be computed and provides a performance reference for other methods

$$P_{hd} = Q\left(2\sqrt{\eta N_p}\right) \qquad (6.58)$$

where

η = the detector quantum efficiency
N_p = the average number of photons per bit

The heterodyne receiver is similar to the homodyne receiver except the local oscillator is not phase locked to the signal and uses a different frequency. The field incident on the photodiode is

$$\vec{E}_r = E_s d_k e^{j\omega_c t} + E_{LO} e^{j\phi} e^{-j\omega_{LO} t} \tag{6.59}$$

where

ϕ = the local oscillator phase

The photodiode current is proportional to the incident power

$$I_d = \alpha P_r = I_{DC} + C E_s E_{LO} d_k e^{j\omega_{IF} t} e^{j\phi} \tag{6.60}$$

where

α and C are constants

$\omega_{IF} = \omega_c - \omega_{LO}$

The photodiode current contains a dc component and an intermediate frequency (IF) containing the PSK modulation. A microwave carrier recovery loop is required to demodulate the PSK signal. This reduces the sensitivity to laser linewidth (requiring 0.2% to 0.4% of bit rate stability) but sacrifices 3 dB in performance

$$P_{het} = Q\left(\sqrt{2\eta N_p}\right) \tag{6.61}$$

Changing the modulation to differential PSK (DPSK), the heterodyne receiver can be simplified, replacing the carrier recovery loop with a one bit period delay and a multiplier to demodulate the signal as shown in Figure 6.33. The bit error performance of this method is

$$P_{DPSK} = \frac{1}{2} e^{-\eta N_p} \tag{6.62}$$

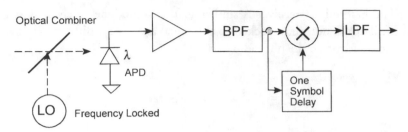

DPSK Hetrodyne Receiver

Figure 6.33 Heterodyne receiver for DPSK.

The theoretical performance of the coherent detectors for optical systems are sumarized in Table 6.6 and the bit error probability is plotted in Figure 6.34.

6.4 Incoherent Demodulation

When a coherent reference is not available, incoherent demodulation must be used. Since coherent demodulation requires a phase reference, incoherent methods do not use signal phase information and suffer a performance degradation. Incoherent demodulation can be used for amplitude, phase and frequency modulation. Since incoherent demodulation pays a price in performance, simpler implementation and lower cost are the main advantages.

Table 6.6
Performance of Coherent Detectors in Optical Systems

Signaling	Receiver	Required Frequency Stability	Bit Error Probability
OOK	direct detection	n.a.	$e^{-2\overline{N}}$
PSK	homodyne	0.02–0.05	$Q\left(2\sqrt{\eta\overline{N}}\right)$
PSK	heterodyne	0.2–0.5	$Q\left(\sqrt{2\eta\overline{N}}\right)$
DPSK	heterodyne	00.3–0.7	$\dfrac{1}{2}e^{-\eta\overline{N}}$

where $= \overline{N}$ = mean photons/bit, η = quantum efficiency

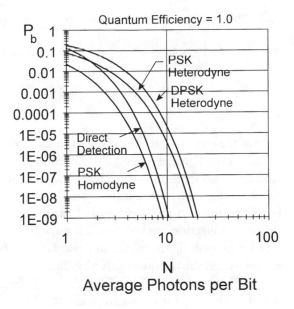

Figure 6.34 Bit error probability for optical receivers.

6.4.1 Amplitude Modulation

The use of discrete tones for digital communication is as old as PCM technology and amplitude modulated tones are still used in supervisory control systems, ripple control systems and optical systems. OOK transmits a "one" using a tone of duration, T, with no tone representing a "zero."

$$y_i(t) = \sqrt{2}A\cos(2\pi f_c t + \theta) \qquad for\ i = 1$$

$$y_i(t) = 0 \qquad\qquad\qquad for\ i = 0 \qquad (6.63)$$

The phase is assumed to be unknown and envelope detection is used to demodulate the tone. The envelope detector can be implemented in several ways and a common theoretical treatment assumes a square law nonlinearity is used to eliminate the phase information. Squaring the signal results in a baseband signal proportional to A^2 and a twice frequency term. Removing the higher frequency terms with a lowpass filter leaves a baseband signal proportional to the signal envelope. The performance in AWGN depends on the setting of the detection threshold. With an optimum threshold, the bit error performance is approximately

$$P_{OOK} \approx \frac{1}{2}e^{-\frac{\gamma}{4}}$$

(6.64)

It has been shown that the error rate at the optimum threshold is primarily due to noise in the absence of the signal exceeding the threshold [3].

The performance of OOK is not just of historical interest. Many optical systems use OOK of the lightwave for transmission. In the case of LEDs and semiconductor lasers, the light output is pulse modulated and direct detection with APDs or homodyne detection can be used. It is difficult to reach the theoretical performance limits with lightwave technology. Using a bit error probability of 10^{-9} as a reference, the quantum detection limit is about 10 photons per bit with homodyne detection of PSK requiring about 18 photons per bit. In practice, the required energy per bit ranges from about 50 to several hundred depending on the bit rate and the laser technology used.

6.4.2 Frequency Modulation

6.4.2.1 Tone Detection

When a coherent carrier reference is unavailable, incoherent detection must be used. The incoherent detection technique utilizes bandpass filters tuned to the frequencies of the individual tones followed by envelope detectors, as shown in Figure 6.35. The outputs of the envelope detectors are sampled at the end of a symbol period and the transmitted tone is taken to be the tone which produces the greatest output at the envelope detector.

Although the analysis is difficult, the bit error probability is remarkably simple

$$P_{FSK} = \frac{1}{2}e^{-\frac{\gamma}{2}}$$

(6.65)

6.4.2.2 Discriminator Detection

The FM discriminator can be used as a detector for FSK signals and the performance approaches that of a coherent matched filter. The basic FM discrimi-

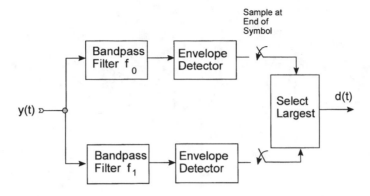

Figure 6.35 Incoherent FSK receiver.

nator is show in Figure 6.36. The input noise is assumed to be AWGN with a one-sided spectral density of N_0. The signal input to the discriminator is:

$$r(t) = \sqrt{2}A\cos(\omega_c t + \theta(t)) + \tilde{n} \qquad (6.66)$$

The bandpass filter output is:

$$y(t) = \sqrt{2}A\cos(\omega_c t + \Omega(\tau)) + \tilde{n}_I \cos(\omega_c t) + \tilde{n}_Q \sin(\omega_c t) \qquad (6.67)$$

where

$$\Omega(t) \approx \int \theta(\tau)h_L(t - \tau)\, d\tau$$
$h_L(t) =$ the equivalent lowpass impulse response of the bandpass filter

The output of the bandpass filter can be represented by a phasor diagram as shown in Figure 6.37. The angle, φ, represents the effective phase error intro-

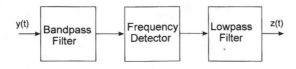

Figure 6.36 FM discriminator.

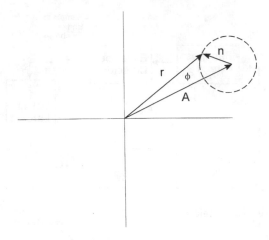

Figure 6.37 Phasor diagram.

duced by the input noise. For the case of no modulation ($\Omega(t) = 0$) and high signal-to-noise ratio, the phase error is:

$$\varphi(t) \approx \frac{\tilde{n}_I}{\sqrt{2}A} \qquad (6.68)$$

The frequency detector output is proportional to the derivative of the phase angle

$$\tilde{w} \approx K_d \frac{d\varphi}{dt} = \frac{K_d \dfrac{d\tilde{n}_I}{dt}}{\sqrt{2}A} \qquad (6.69)$$

The noise spectral density at the frequency detector is then

$$S_{ww} \approx \frac{K_d^2}{\sqrt{2}A^2}\omega^2 N_0, |f| \leq \frac{B_{IF}}{2} \qquad (6.70)$$

$$S_{ww} = 0, |f| > \frac{B_{IF}}{2}$$

At low signal-to-noise ratios, Rice [46] argues that the noise at the output of the detector can be approximated by the rising Gaussian noise summed with impulsive noise consisting of "clicks," each having an area of 2π radians. The rate at which the clicks occur depends upon the signal-to-noise ratio and the bandpass filter bandwidth. Two types of detection filters are considered, a reset integrator (RSI) and a Bessel, linear phase filter whose output is sampled at the peak signal point.

The mean square noise output of the RSI due to the rising noise component is

$$\sigma^2 = \int_{-\infty}^{+\infty} S_{ww} \left| H(j\omega) \right|^2 d\omega \qquad (6.71)$$

where

$H(j\omega)$ = the transfer function of the RSI

With some elementary trigonometry and calculus, the mean square noise can be evaluated and found to be

$$\sigma^2 = \frac{2\pi K_d^2 N_0}{E_b} \left[B_{IF}T - \frac{1}{\pi} \sin \pi B_{IF}T \right] \qquad (6.72)$$

The peak signal at the output of the RSI is

$$z_p = \pi K_d \alpha \, \Delta f \, T \qquad (6.73)$$

where

Δf = the peak-to-peak frequency deviation

The factor, α, is included to take into account the filtering effect of the bandpass filter. For an infinitely wide bandpass filter, $\alpha = 1.0$. For a narrower filter, $0 < \alpha < 1$. The peak signal-to-noise ratio at the RSI output is

$$\frac{z_p}{\sigma} = \alpha \sqrt{\frac{\pi}{2}} \Delta f \, T \sqrt{\frac{E_b}{N_0} \frac{1}{f(B_{IF}T)}} \qquad (6.74)$$

where

$$f(B_{IF}T) = B_{IF}T - \frac{1}{\pi}\sin \pi B_{IF}T$$

The probability of error due to the rising noise component is

$$P_1 = Q\left(\frac{z_p}{\sigma}\right) = Q\left(1.25C_1 \frac{\Delta f\, T}{\sqrt{B_{IF}T}}\sqrt{\gamma}\right) \tag{6.75}$$

where

$$C_1 = \frac{\alpha}{\sqrt{1 - \dfrac{\sin \pi B_{IF}T}{\pi B_{IF}T}}}$$

The function, C_1, is approximately 1 for $B_{IF}T > 1$. When the IF bandwidth is less than the symbol rate, C_1 increases rapidly.

The "click" noise component may be considered to be an impulse with area, 2π. If a click occurs during a symbol period, the RSI output will be

$$z_c = 2\pi K_d \tag{6.76}$$

If the click is of opposite polarity to the signal, an error will be made if:

$$\left|z_c\right| \geq \left|z_p\right| \tag{6.77}$$

which leads to

$$\Delta f T \leq 2$$

This includes the general range of deviations of interest. Rice has shown that the average number of clicks per second is approximately

$$N_{clicks} \approx \frac{\Delta f}{2} e^{-\frac{1}{B_{IF}T}\gamma} \tag{6.78}$$

The average error probability due to the clicks is approximately

$$P_2 \approx \frac{\Delta f T}{2} e^{-\frac{1}{B_{IF}T}\gamma} \tag{6.79}$$

The total bit error probability for RSI detection is the sum of the errors due to the Gaussian component and the click component

$$P_{RSI} \approx Q\left(1.25 \frac{\Delta f T}{\sqrt{B_{IF}T}}\sqrt{\gamma}\right) + \frac{\Delta f T}{2} e^{-\frac{1}{B_{IF}T}\gamma} \tag{6.80}$$

for $B_{IF}T > 1$

The technique employed for evaluating the bit error rate using filter/sample (F/S) detection follows the same steps as the RSI analysis but is complicated by the necessity to evaluate specific filter responses. The analysis of errors due to clicks is more complex for F/S detection than for RSI since the actual waveshape of the filter impulse response must be evaluated. Using an analysis like that of Schilling, et al.[47], the bit error probability F/S detection is approximately

$$P_{FS} \approx Q\left(C_2 \frac{\Delta f T}{(B_{IF}T)^{\frac{3}{2}}}\sqrt{\gamma}\right) + \frac{\Delta f T}{2\pi} e^{-\frac{1}{B_{IF}T}\gamma} \tag{6.81}$$

where

$$C_2 = 0.345\beta\left(\frac{B_{IF}}{B_{LP}}\right)^{\frac{3}{2}}$$

The constant, $\beta < 1$, depends on the filtering of the combined bandpass and lowpass filters. For the sake of comparison, β is assumed to be one and the lowpass filter bandwidth is one half the bandpass filter bandwidth. With these assumptions, C_2 is approximately one and the performance of the F/S detector is similar to the RSI detector. A more complete analysis shows that the optimum performance is obtained when the modulation index ($\Delta f T$) is about 0.7 and the performance is close to that of the coherent FSK matched filter.

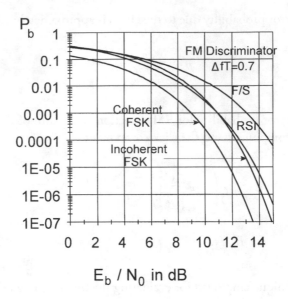

Figure 6.38 Incoherent FSK bit error probability.

The previous results can be extended to the case of M-ary (or multilevel) FSK. The asymptotic behavior of the FM discriminator detector for M-ary FSK is of the form

$$P_{MFSK} \approx \frac{M-1}{2} \Delta f\, T\, e^{-\frac{1}{\Delta f\, T}\frac{\gamma}{M}} \qquad (6.82)$$

The bit error performance degrades 3 dB each time the number of levels doubles. This is in marked contrast to the optimum incoherent matched filter detector whose performance is asymptotically the same for any number of levels. The bit error probability for FM discriminator detection is plotted in Figure 6.38 for typical values of the constants.

6.5 Bit Error Performance with Intersymbol Interference and Fading

To this point, the bit error performance has only been considered for the case of AWGN or shot noise, in the case of optical receivers. Real systems have to operate in environments with all types of distortion, interference, and fading. The most common distortion is linear filtering by components in the channel and the system. Filtering causes intersymbol interference with energy from one symbol spreading to adjacent symbols, degrading symbol decision margins. Interference ranges from intermodulation distortion products and crosstalk to non-Gaussian noise from outside sources. Fading is considered to be a multiplicative noise causing the signal level to fluctuate. In radio systems, fading can be categorized as "slow" or "fast" (relative to the signaling rate), and as specular or diffuse multipath. In some cases, fading is caused by environment conditions which alter the propagation characteristics. System performance due to intersymbol interference will be considered first.

6.5.1 Intersymbol Interference

The PCM is a sampled system with a sampling at the symbol rate. The system can be modeled as a discrete time system as shown in Figure 6.39. The equivalent system response is modeled as a discrete time, finite impulse response (FIR) filter with $2N+1$ taps. The sampled response is

$$z_k = \sum_{j=-N}^{j=+N} d_j h_{k-j} + \tilde{w}_k \tag{6.83}$$

The noise samples are assumed to have a Gaussian distribution but are not necessarily white. If the input is a unit signal, the output signal is

$$z_k = h_0 d_k + \sum_{j \neq k} d_j h_{k-j} + \tilde{w}_k \tag{6.84}$$

The first term is the desired signal, the second term is intersymbol interference for preceding and succeeding symbols, and the last term is the noise.

Without loss of generality, the data sequence, d, will be assumed to be random taking on values ± 1. There will be some data sequence which makes the intersymbol interference term take on a maximum value

$$\Delta z = \sum_{j \neq k} \left| h_{k-j} \right| \tag{6.85}$$

When this occurs, the symbol decision margin is degraded by

$$\Delta m = 20 \log \left(1 - \frac{\Delta z}{h_0} \right) in \ dB \tag{6.86}$$

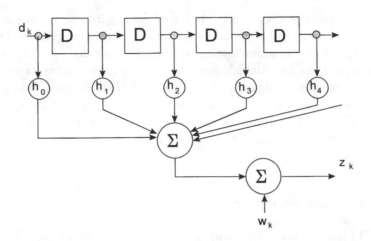

Figure 6.39 Discrete model.

This is equivalent to using the worst case eye opening to estimate the performance degradation due to intersymbol interference. The intersymbol interference component can be added to the noise as a composite noise component

$$\tilde{n}_t = \tilde{v}_{ISI} + \tilde{w}$$

$$\tilde{v}_{ISI} = \sum_{j \neq k} d_j h_{k-j} \tag{6.87}$$

For a binary random sequence, the bit error probability is the probability that $|z_k|$ exceeds h_0. For no intersymbol interference and Gaussian noise this has been shown to be

$$P_b = Q\left(\frac{h_0}{\sigma}\right) \tag{6.88}$$

With intersymbol interference, the equivalent noise is the sum of the Gaussian noise and the intersymbol interference term. The bit error probability can be expressed in terms of the characteristic function of the noise

$$P_b = \frac{1}{2} - \frac{h_0}{\pi} \int_0^\infty \frac{\sin h_0 \lambda}{h_0 \lambda} F(\lambda) \, d\lambda \tag{6.89}$$

where

$F(\lambda)$ = the characteristic function of the noise

If the random noise is assumed to be statistically independent of the intersymbol interference, the characteristic function of the sum of the two noises is the product of the individual characteristic functions

$$F(\lambda) = F_{ISI}(\lambda)F_w(\lambda) \tag{6.90}$$

The characteristic function of Gaussian noise is

$$F_w(\lambda) = e^{-\frac{\lambda^2 \sigma^2}{2}} \tag{6.91}$$

For a binary data sequence, the intersymbol interference is the sum of random variable taking on values, $h_j = \pm 1$ with equal probability. Thus each term has a probability density function

$$p_h(x) = \frac{1}{2}\delta(x - h_j) + \frac{1}{2}\delta(x + h_j) \tag{6.92}$$

The characteristic function, F_{ISI}, is computed to be

$$F_{ISI} = \prod_{j \neq 0} \cos h_j \lambda \tag{6.93}$$

Although the bit error probability can be computed using these equations, better insight can be gained, and numerical computational efficiency improved, by separating the bit error probability calculation into the sum of the error due to the Gaussian noise and a term dependent on the intersymbol interference

$$P_b = P_g + P_{ISI}$$

$$P_b = Q\left(\frac{h_0}{\sigma}\right) + \frac{h_0}{\pi}\int_0^\infty \frac{\sin h_0 \lambda}{h_0 \lambda}\left[1 - \prod_{j \neq 0} \cos h_j \lambda\right] e^{-\frac{\lambda^2 \sigma^2}{2}} \, d\lambda \tag{6.94}$$

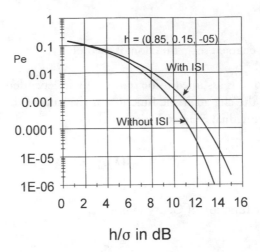

Figure 6.40 Probability of bit error for an example with intersymbol interference.

Despite its imposing appearance, the integral containing the intersymbol interference term can be numerically integrated using a programmable calculator or
an engineering mathematics program. As an example, a filtered baseband NRZ
signal has a sampled impulse response

$$h_0 = 0.85, \quad h_1 = 0.15, \quad h_2 = -0.05$$

The computed bit error probability is shown in Figure 6.40 with both the total
error probability and the error probability due only to Gaussian noise. The difference between the curves is the contribution of the intersymbol interference
terms. From the figure, the degradation is about 2 dB while the worst case approximation gives an estimate of 3 dB.

Filtering is often done deliberately to bandlimit the PCM signal spectrum.
In this case, the designer makes a tradeoff between the intersymbol interference
degradation and the suppression of out-of-band signal energy. Although the
theoretical performance of bandlimited signals could be computed, there are
experimental data on a number of commonly used PCM signals. The empirical
results offer insight into the relative performance degradation which might be
expected. Figure 6.41 shows the effects of filtering NRZ signals with a third
order Butterworth filter with normalized bandwidths from BT = 0.5 to BT =
1.0. A degradation of 2.5 to 3.0 dB relative to theoretical is typical when the
signal is limited to a bandwidth equal to the bit rate. Of that degradation, about

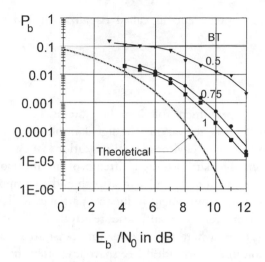

Figure 6.41 Effect of filtering on NRZ PCM.

1 dB is due to the receiver implementation. Figure 6.42 shows similar data for modulated carriers using PSK.

Transmission channels introduce unwanted filtering and the designer must either design the signaling to minimize intersymbol interference or must process the received signal to compensate for the distortion. Compensating

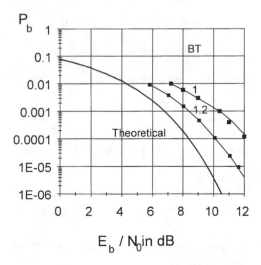

Figure 6.42 Effect of filtering on PSK.

for the intersymbol interference is called equalization and will be discussed later.

6.5.2 Fading

Signal fading plays a prominent role in radio systems. Fading is generally considered a multiplicative factor varying the signal amplitude. In a radio environment, fading can be due to environmental conditions (such as rain) or to path geometry variations. The latter fading is often broken into two parts, a component which is due to a strong specular multipath and a component due to diffuse multipath. The first component is due to a partial cancellation of the direct signal by a single strong reflection and varies relatively slowly as the path geometry changes. In systems where the receiver (or transmitter) can have random siting, the variation in loss over ideal free space loss often includes the specular multipath component along with other undefined losses and associates a probability distribution to the composite loss. Empirical data suggests a lognormal distribution for this factor so that the slowly varying loss has a normal distribution when expressed in dB.

The second component of fading is due to diffuse scattering of the transmitted energy from a variety of reflection points within the path geometry. A reasonable model for this component assumes many multipath signals with random amplitudes and phases resulting in a Rayleigh probability density function

$$p(r) = \frac{r}{\sigma^2} e^{-\frac{r^2}{2\sigma^2}} \; for \; r \geq 0 \tag{6.95}$$

If the Rayleigh fading is slow with respect to the bit rate, the average bit error probability is

$$P_{fading} = \int_0^\infty P_b(\gamma) p(\gamma) d\gamma \tag{6.96}$$

where

γ = the signal-to-noise ratio expressed as the energy per bit to noise spectral density

$p(\gamma)$ = the probability density of the fading

When the envelope has a Rayleigh density, the probability density function of the signal-to-noise ratio, γ, is particularly simple

$$p(\gamma) = \frac{1}{m} e^{-\frac{\gamma}{m}} \qquad (6.97)$$

where

$m =$ the mean signal-to-noise ratio

The bit error probability of the signals which have been discussed are either a function of $Q(\gamma)$, $exp(-\gamma)$ or a combination of these functions (PCM/FM, for example). Taking the exponential case first, the average bit error probability in Rayleigh fading is

$$P = \int_0^\infty \frac{1}{m} e^{-\frac{\gamma}{m}} c_1 e^{-c_2 \gamma} d\gamma \qquad (6.98)$$

This is readily integrated to

$$P = \frac{c_1}{1 + c_2 m} \qquad (6.99)$$

For noncoherent FSK, c_1, $c_2 = 1/2$, so the average bit error probability is

$$P_{NCFSK} = \frac{1}{2 + \bar{\gamma}} \qquad (6.100)$$

where

$m = \bar{\gamma}$, the average E_b/N_0

When the bit error probability is a function of the Gaussian normal function, $Q(x)$, the average bit error probability with Rayleigh fading is

$$P = \int_0^\infty \frac{1}{m} e^{-\frac{\gamma}{m}} c_1 Q\left(\sqrt{c_2 \gamma}\right) d\gamma \qquad (6.101)$$

This can be simplified to

$$P = \frac{c_1}{2m} \int_0^\infty e^{-\beta\gamma} erfc\left(\sqrt{\alpha\gamma}\right) d\gamma \tag{6.102}$$

The integral has a closed form solution [48] and the average bit error probability is

$$P = c_1 \left[1 - \sqrt{\frac{c_2 \bar{\gamma}}{2 + c_2 \bar{\gamma}}} \right] \tag{6.103}$$

For the case of PSK , $c_1 = 1$, $c_2 = 2.0$,

$$P_{PSK} = \frac{1}{2} \left[1 - \sqrt{\frac{\bar{\gamma}}{1 + \bar{\gamma}}} \right] \tag{6.104}$$

Table 6.7 summarizes the bit error probability with Rayleigh fading and Figures 6.43 and 6.44 plot the probabilities for a number of the signaling techniques. These are the systems designer's worst nightmare. The bit error probability is inversely proportional to the signal-to-noise ratio decreasing by a factor of 10 for every 10 dB increase in signal-to-noise ratio implying 50 to 60 dB to achieve a 10^{-6} bit error rate. A more realistic view considers the probability that the signal-to-noise ratio is less than the mean for a given value. Integrating the probability density function, the probability that the signal-to-noise ratio is less than a value, γ, is

$$P(\gamma \leq \bar{\gamma}) = 1 - e^{-\frac{\gamma}{\bar{\gamma}}} \tag{6.105}$$

This function is plotted in Figure 6.45 as a function of the signal-to-noise ratio relative to the mean signal-to-noise ratio expressed in dB. About 10% of the time, the signal-to-noise ratio is more than 10 dB below the mean. The signal-to-noise ratio is more than 20 dB below the mean about 1% of the time.

6.6 Error Correction

Error correction is becoming increasingly common in PCM systems. Many of the first error correction systems were simply overlaid on existing uncoded sys-

Table 6.7
Bit Error Probability with Rayleigh Fading

Signaling	Bit Error Probability with Rayleigh Fading
NRZ, Biphase, Raised Cosine	$\frac{1}{2}\left[1-\sqrt{\dfrac{\overline{\gamma}}{1+\overline{\gamma}}}\right]$
Quaternary Raised Cosine	$\frac{1}{2}\left[1-\sqrt{\dfrac{\overline{\gamma}}{1.25+\overline{\gamma}}}\right]$
AMI (100% Pulse Width)	$\frac{1}{2}\left[1-\sqrt{\dfrac{\overline{\gamma}}{2+\overline{\gamma}}}\right]$
Partial Response, Class 1	$\frac{1}{2}\left[1-\sqrt{\dfrac{\overline{\gamma}}{\frac{16}{\pi^2}+\overline{\gamma}}}\right]$
Partial Response, Class 4	$\frac{1}{2}\left[1-\sqrt{\dfrac{\overline{\gamma}}{4+\overline{\gamma}}}\right]$
PSK, QPSK	$\frac{1}{2}\left[1-\sqrt{\dfrac{\overline{\gamma}}{1+\overline{\gamma}}}\right]$
M-ary PSK, M > 4	$\left[1-\sqrt{\dfrac{\overline{\gamma}}{\frac{1}{\sin\frac{\pi}{M}}+\overline{\gamma}}}\right]$
MSK	$\left[1-\sqrt{\dfrac{\overline{\gamma}}{1+\overline{\gamma}}}\right]$
Incoherent FSK	$\dfrac{1}{2+\overline{\gamma}}$

tems. This is possible by inserting an encoder prior to the symbol encoder and a decoder after the symbol detector as shown in Figure 6.46. The communication channel can be considered to be a binary symmetric channel (BSC) and the performance improvement can be computed from the known code characteristics. For example, if a block code with a length of 20 bits is used, which can

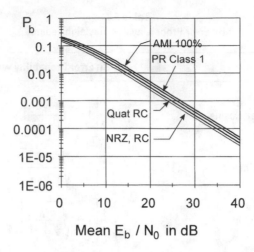

Figure 6.43 Bit error probability for Rayleigh fading of baseband signals.

correct all double errors, the probability of an uncorrected error is the probability that three or more errors occur in 20 bits. To a first approximation, the probability of an uncorrected error is about

$$P_{be} = \sum_{t+1}^{\infty} \frac{N!}{(N-j)!\,j!} p^j (1-p)^{N-j} \approx \frac{N!}{(N-t-1)!\,(t+1)!} p^{(t+1)} \quad (6.106)$$

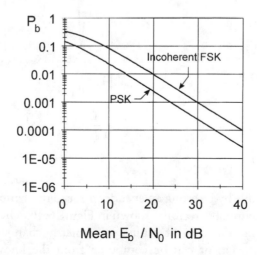

Figure 6.44 Bit error probability for Rayleigh fading of modulated carrier signals.

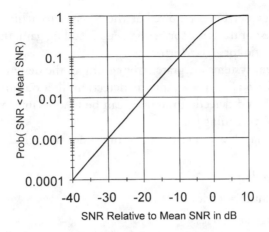

Figure 6.45 Probability that a SNR is less than the mean SNR.

where

P_{be} = the probability of at least one error in the block of N bits
p = the bit error probability
t = the number of correctable errors

A 10^{-3} raw error rate would be corrected to a block error rate of about 10^{-6}. If there are k information bits per block, the decoded information error rate would be reduced by an additional factor approximately equal to k. While this may seem to be a big improvement, the code rate is higher than the uncoded rate and

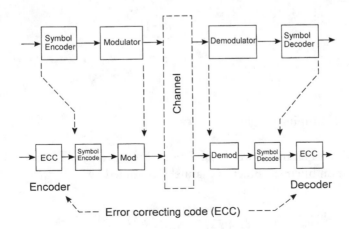

Figure 6.46 Adding error correction to an existing system.

additional energy per bit is required at the receiver to achieve the 10^{-6} error rate. The true test of the error correcting code is coding gain, the net decrease in signal-to-noise ratio for a given error rate.

More recent systems integrate error coding in the design, combining coding with modulation. By quantizing the detection filter output (soft decisions), maximum likelihood detection methods can be used to improve the error correction and increase coding gain.

Three general classes of error correcting codes are considered

- Block codes;
- Convolutional codes;
- Concatenated codes.

For each of these types of codes, there are several decoding methods with differing performance. In this section, the error correction performance is considered for the more popular coding methods.

6.6.1 Block Codes

The error correcting performance of a block code is determined by the Hamming weight distribution of the code. The Hamming weight is the number of places which differ between two code words. The Hamming weight is the "distance" between code vectors and the maximum likelihood decoding chooses the code word which is closest to the received word. A given code has a weight distribution which gives the number of codes with a given weight. The minimum weight determines the error correcting capability

$$N_{ce} = \frac{w-1}{2} \tag{6.107}$$

where

$w =$ Hamming weight

A code with a Hamming weight of 7 can correct 3 errors. The probability of a block error on binary symmetric channel is bounded by

$$P_{be} \leq \sum_{m}^{N} \left(\frac{N!}{(N-j)!\,j!} \right) p^j (1-p)^{N-j} \tag{6.108}$$

where

$$m = \frac{w_{\min} + 1}{2}, \qquad w_{\min} \, odd$$

$$m = \frac{w_{\min}}{2}, \qquad w_{\min} \, even$$

$w_{\min}$ = the minimum Hamming distance

The bound is exact for codes which meet the sphere-packed criteria such as the single error correcting Hamming codes. This is also a good practical estimate for all block codes.

The bit error probability is less than the block error probability, and is the average number of information bit errors per block times the block error probability.

$$P_b = \frac{\text{ave bit errors}}{\text{block}} P_{be} \approx \frac{t+1}{k} P_{be} \tag{6.109}$$

If a code corrects t errors, when a block error occurs $t + 1$ errors are most likely. As an example, the Golay (23,12) code has a Hamming distance of 7 and corrects 3 errors. The bit error rate is approximately 1/3 the block error probability.

The bit error probability for the BSC with maximum likelihood decoding is approximated by the first term in the series for error rates greater than about 10^{-3}.

$$P_b \approx \frac{t+1}{k} \frac{N!}{(N-t-1)!(t+1)!} p^{t+1} \tag{6.110}$$

If the channel is assumed to be AWGN and the received signal is processed in a maximum likelihood detector, the block error probability is bounded by the union bound

$$P_{be} \leq \sum_{w \in W} N_w Q\left(\sqrt{\frac{2wrE_b}{N_0}}\right) \tag{6.111}$$

where

> W is the set of code weights
> N_w is the number of codes of a given weight
> $r =$ the code rate

Although this is only an upper bound, it has been found to be reasonably tight for error rates greater than about 10^{-2} (good enough for government work). Berlekamp [49] has developed a tangential union bound which extends the bound to error rates less than 10^{-2} for those cases where very low rates are important. A further approximation, useful for systems design, uses only the first term in the union bound series.

$$P_{be} \approx N_{w_{min}} Q\left(\sqrt{\frac{2w_{min} r E_b}{N_0}}\right) \tag{6.112}$$

Assuming the channel signaling uses optimum antipodal symbols (such as NRZ-L or PSK), the uncoded error rate is given by

$$P_e = Q\left(\sqrt{\frac{2E_b}{N_0}}\right) \tag{6.113}$$

Neglecting the multiplicative constant and comparing the signal-to-noise ratio terms, the asymptotic coding gain for the AWGN channel with maximum likelihood decoding is

$$G_{AWGN} = 10\log w_{min} r \tag{6.114}$$

On a BSC with maximum likelihood decoding, the asymptotic coding gain is

$$G_{BSC} = 10\log \frac{r(w_{min} + 1)}{2} \tag{6.115}$$

Thus the "soft decision" decoding gain is about 3 dB greater than hard decision decoding. The asymptotic coding gain for some common block codes is listed in Table 6.8.

Table 6.8
Coding Gains of Some Block Codes

Block Code	Rate	Minimum Weight	Coding Gain in dB	
			BSC	AWGN
(18,9)	0.5	6	2.43	4.77
(16,11)	0.6875	4	2.35	4.4
(24,12) Golay	0.5	8	3.52	6.02
(32,21)	0.6563	6	3.11	5.95

The approximate bit error probability for some typical block codes on BSC and AWGN channels is shown in Figures 6.47 and 6.48, assuming maximum likelihood decoding.

6.6.2 Convolutional Codes

Convolutional codes are typically decoded using either syndrome decoders or Viterbi decoders. The syndrome decoder is most easily understood by using a 1/2 rate systematic code as an example. By definition, one of the outputs of a

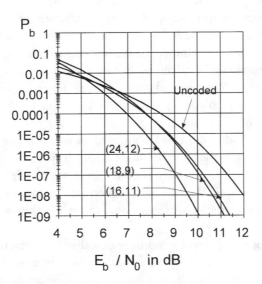

Figure 6.47 Bit error probability for block codes on BSC.

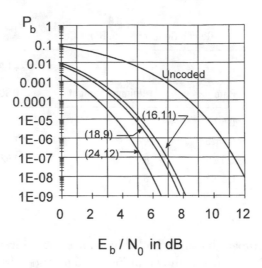

Figure 6.48 Bit error probability for block codes on a AWGN channel.

systematic code is the input data stream so a 1/2 rate systematic code output consists of the input data bits alternating with parity bits. The transmission channel is assumed to be two BSCs with the uncoded data bits on one channel and the parity bits on the second channel, as illustrated in Figure 6.49. At the receiver, the data bit stream is fed to a replica of the encoder and the parity bits produced are compared with the received parity bits. The result is called the syndrome sequence, or just the syndrome. If there are no channel errors, the syndrome is zero. A nonzero syndrome indicates channel errors which must be corrected. If the code constraint length is short enough, the syndromes resulting from error sequences can be exhaustively computed and the minimum distance sequence determined. The received syndrome can address a lookup table which provides an output which corrects the received bit. The bit decision is fed back to the syndrome computation.

The syndrome decoder can be thought of as a "sliding block" decoder. With the 1/2 rate code, the effective block length is twice the constraint length. A constraint length, 11, systematic code has been found with a Hamming distance of 7 and, consequently, will correct 3 errors. Treated as a block code, the decoded bit error probability is proportional to the fourth power of the channel bit error.

Convolutional codes with syndrome decoding are attractive because of the relatively simple encoder and decoder implementation as well as the ability to interleave the code to correct burst errors. The codes can be interleaved by inserting fixed shift registers between each register in the encoder. If the shift reg-

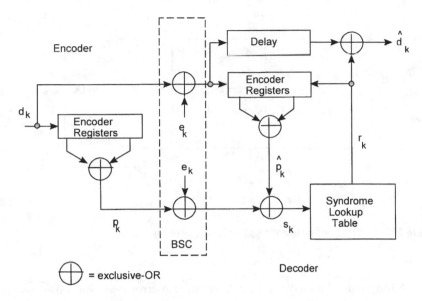

Figure 6.49 Syndrome decoder for 1/2 rate systematic code.

ister length is L bits, a noise burst of up to L bits will only affect one tap in the decoder and the decoder is able to correct bursts up to tL long, where t is the number of correctable errors. Interleaving can be very powerful on fading channels where errors tend to occur in bursts.

Convolutional codes are most often associated with Viterbi decoding. To fully appreciate the ideas behind Viterbi decoding and to understand the decoding performance, the constraint length, 3, nonsystematic code shown in Chapter 5 and described by Viterbi and Omura [11] will be used as an example. A code with a constraint length, K, has $K-1$ registers, in the example, the encoder has two registers as shown in Figure 6.50.

The encoder can be completely described by the state of the registers. A constraint, K, code is completely described by 2^{K-1} states. For the example, the encoder has four states as illustrated in Figure 6.51. The diagram is known as a trellis diagram showing the states of the encoder and the possible transitions between states. The nomenclature used by Viterbi will also be used here. The states are labeled with subscripts corresponding to the decimal equivalent of the state binary value. The least significant bit of the state represents the most recent bit into the encoder. The branches are labeled with the output bits of the encoder with the left hand bit corresponding to the topmost output. Transitions caused by a "one" input are shown by a dashed line.

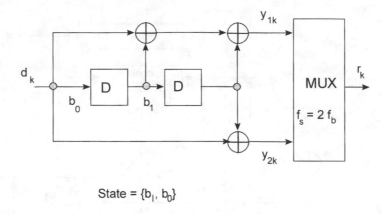

State = {b_1, b_0}

Figure 6.50 Half rate nonsystematic code with constraint length, $K = 3$.

Suppose the data sequence is all "zeros." An error event can cause the state transitions to take a divergent path from the all "zeros" path. One such path goes from state s_0 to s_1 to s_2 and finally back to s_0. The Hamming distance for this path is the sum of the weights associated with each branch, in this case, a total distance of 5. Examining the trellis shows that this is the minimum distance path and, hence, this code has a minimum distance of 5 with a 2 error correcting capability. The weight of other paths can be investigated to deter-

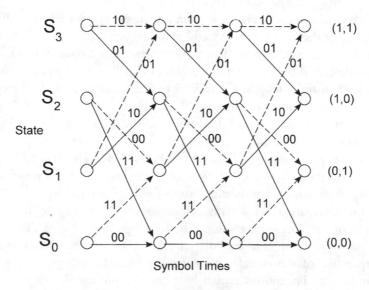

Figure 6.51 Trellis diagram for an example of nonsystematic convolutional code.

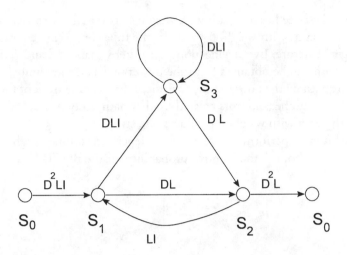

Figure 6.52 State diagram for an example of convolutional code.

mine the weight structure of the code. Because of the periodic structure of the trellis diagram, an alternative state diagram can be drawn which will provide a direct way of evaluating the weight structure of convolutional codes. The state transitions are shown in a state diagram, Figure 6.52, in which the s_0 state is split apart. In this diagram the branches are labeled using a variable, D^w, to indicate the weight (w) of the branch, a label, L, to count the number of times a branch has been traversed and a variable, I, to indicate if the branch was caused by a "one." The branch from s_1 to s_3 in the example has a weight of 1 and was caused by a "one."

Signal flow graph methods can be used to compute a transfer function, $T(D)$, from input (state s_0) to output (also s_0). In this case the transfer function is computed to be

$$T(D,L,I) = \frac{D^5 L^3 I}{1 - DLI(1+L)} \tag{6.116}$$

First, ignore L and I by setting them equal to 1 and expand the remaining fraction into a power series in the variable, D.

$$T(D) = \frac{D^5}{1 - 2D} = D^5 + 2D^6 + \cdots + 2^i D^{5+i} \tag{6.117}$$

This shows the code has one path with weight 5, two paths with weight 6, and, in general, 2^i paths with weight $5+i$. Thus the function, $T(D)$, enumerates the code weight structure. By carrying along the other variables, L and I, the length of a given path can be obtained from the power of L and the number of "ones" creating the path by the power of the variable, I. By keeping track of the number of "ones," the number of errors created by that path is obtained. Again for this example, the path with weight 5 produces one error.

The bit error performance of a code is dependent on the weight structure. Using the union bound, the bit error probability for the BSC [11] is

$$P_{BSC} \leq \frac{1}{b} \frac{\partial T(D,I)}{\partial I}\bigg|_{I=1,D=\sqrt{4p(1-p)}} \tag{6.118}$$

where

the code rate is b/n
p = input bit error probability

For the example code, the bit error probability is

$$P_{BSC} \leq \left(4p(1-p)\right)^{\frac{5}{2}} + 2\left(4p(1-p)\right)^3 + \cdots$$

$$P_{BSC} \approx 32p^{2.5} \tag{6.119}$$

The bit error performance for an AWGN channel assumes "soft" decisions are made on the received symbols. An upper bound on the AWGN bit error probability is

$$P_b = \frac{1}{b} Q\left(\sqrt{\frac{2rd_f E_b}{N_0}}\right) e^{rd_f \frac{E_b}{N_0}} \frac{\partial T(D,I)}{\partial I}\bigg|_{I=1,D=e^{-r\frac{E_b}{N_0}}} \tag{6.120}$$

where

r = the code rate
d_f = the minimum "free" distance (the minimum weight)

The bit error probability can be approximated by the first term in the series

$$P_b \approx \frac{1}{b} Q\left(\sqrt{\frac{2rd_f E_b}{N_0}} \right)$$ (6.121)

As in the case of block codes, the asymptotic coding gain is given by rd_f. The minimum free distance and asymptotic coding gain of some 1/2 and 1/3 rate codes are summarized in Table 6.9.

A maximum likelihood decoder is required to achieve the theoretical bit error performance predicted by the union bounds. The Viterbi decoder uses an algorithm which can meet this goal. The Viterbi algorithm (VA) can be explained using the trellis diagram. In the example code, there are two branches entering each state. For a given state, the VA keeps track of two parameters, an accumulated metric proportional to the loglikelihood and the branch which has the maximum loglikehood. At each symbol time, the metric is updated for each branch entering a state and the branch with the greatest loglikehood is marked. The path leading to the state with the greatest loglikehood is followed back a

Table 6.9
Coding Gains of 1/2 and 1/3 Rate Codes

Constraint Length	Code Rate	Systematic Codes		Nonsystematic Codes	
		d_f	Coding Gain in dB	d_f	Coding Gain in dB
2	½	3	1.76	3	1.76
3	½	4	3.0	5	3.98
4	½	4	3.0	6	4.77
5	½	5	3.98	7	5.44
6	½	6	4.77	8	6.0
7	½	6	4.77	10	6.99
8	½	7	5.44	10	6.99
2	⅓	5	2.22	5	2.22
3	⅓	6	3.0	8	4.26
4	⅓	8	4.26	10	5.23
5	⅓	9	4.77	12	6.0
6	⅓	10	5.23	13	6.37
7	⅓	12	6.0	15	6.99
8	⅓	12	6.0	16	7.27

number of symbol times and the bit which started the path is output as the correct decision. The loglikelihood metric measures the distance between the path and the received sequence and is largest when the distance is smallest. Theoretically, one would have to trace the path all the way back to the first symbol. In practice, the path is truncated after a number of symbols, typically 2 to 5 times the code constraint length.

The VA decoder needs storage for about $5 \times L \times 2^L$ parameters and 2^{L-1} computational units for a code with a constraint length, L. The regular structure of the VA lends itself to implementation in VLSI circuits. Quantization of the received signal to as little as 4 bits per symbol can be used without degrading the AWGN performance by more than about 0.25 dB.

6.6.3 Concatenated Codes

To achieve large coding gains either requires long block codes or convolutional codes with large constraint lengths. An alternative to single, long codes, concatenating two short codes can effectively provide the performance of a long code. Concatenated codes have been used on many deep space vehicles where communication efficiency is extremely important. In these programs, a block code has been concatenated with a convolutional code. In recording applications, two block codes (Reed-Solomon) are concatenated to provide powerful burst error correction reducing a raw bit error rate from about 10^{-5} to less than 10^{-8}. Most recently, a new class of concatenated codes, known as the "turbo" codes, have been introduced which have astonishing performance, approaching the Shannon bound [50]. The use and performance of concatenated codes is very specific to the application and the discussion will be limited to a few specific examples.

6.6.3.1 Concatenated Reed-Solomon Codes

The Reed-Solomon codes are a class of cyclic, nonbinary block codes. Each symbol in the R-S code is a binary word with q bits and the block length is

$$N_{RS} = 2^q - 1 \tag{6.122}$$

A R-S code using 4 bit symbols has a block length of 15. An additional parity symbol can be added to extend the block length by one symbol. Since each symbol is 2^q bits long, a t symbol error correcting R-S code can correct t bursts of 2^q bits. If the extended R-S code has r parity symbols, it is able to correct t symbol errors and s symbol erasures for

$$2t + s \leq r + 1 \tag{6.123}$$

The R-S code is often used as the outer code in a concatenated coding method. Forney observed that the purpose of the inner code is to reduce the error rate to a moderate value ($\sim 10^{-3}$) and the outer code to then reduce the error rate to the final desired rate [17]. The rate of the inner code is chosen to be somewhat higher than the desired overall rate with the outer code rate reducing the composite rate. Forney gives a number of examples of concatenated codes which reduce a raw error rate from 0.01 to less than 10^{-12}. One such example uses a (15,11) BCH block code as the inner code which corrects single errors concatenated with a (76,52) R-S outer code. A number of space communication links use convolutional codes as the inner code with R-S outer codes. High data rate, high density magnetic recorders use a concatenated code with R-S codes used for both the inner and outer codes.

When the R-S code is used as the outer code, the bit error probability as a function of the input symbol error probability is

$$P_e = \sum_{j>t} \frac{1}{2} \frac{j}{N-1} \binom{N}{j} p^j (1-p)^{N-j} \tag{6.124}$$

Berlekamp has tabulated this function for a number of extended R-S codes in terms of the input error rate for a given output error rate. As an example, a (16,14) R-S code with 4 bit symbols can correct an input symbol error rate of 2.6×10^{-4} to an output error rate of 10^{-9}.

The performance of the examples is typical of applications using concatenated codes. The real "trick" in these applications is to get the inner code to do its job. The coding gain decreases with increasing error rate and at very poor error rates, the inner decoder may fail and actually produce more output errors than input errors. This can occur on channels with very deep fades or with highly non-Gaussian noise. With sophisticated coding systems, the overall performance of the system is binary, it either works with no errors or is totally useless.

6.6.3.2 Turbo Codes

Turbo codes are a recent variation of concatenated codes which offer impressive performance. Whereas the previous concatenated codes have been in series (an inner and an outer code), turbo codes are in parallel. Although the development of the turbo codes is still proceeding, the basic encoder, described in Chapter 5, consists of two convolutional encoders in parallel, each producing a parity bit stream. One encoder is fed directly by the data stream while the second encoder is fed by an interleaved data stream. In the basic configuration the code rate is

1/3 but other rates are obtained by "puncturing" (selectively removing parity bits) the code.

The first realizations of the turbo codes used recursive systematic convolutional (RSC) codes which have feedback connections as well as the conventional feed forward taps. Unlike conventional convolution encoders, the RSC encoder can produce infinitely long sequences from a low weight input. At the same time, some low weight sequences can also be produced. This is where the interleaver comes in. Interleaving the second encoder can break up those input sequences which lead to low weight outputs of the code. The weight is the sum of the weights of the outputs of the code and depends on the combination of the RSC encoders and the interleaving.

The decoding of turbo codes uses a novel iterative scheme with a basic decoder module as shown in Figure 6.53. The decoder can be implemented as a "pipeline" combination of the basic module or by using one decoder module and recycling the decoded results back through the same module. The decoding performance using the iterative decoding is quite amazing. Figure 6.54 shows the required signal-to-noise ratio as a function of the number of iterations for a 1/2 rate turbo code with 256×256 interleaving at a bit error rate of 10^{-5}. After 18 iterations, the performance is within 0.7 dB of the Shannon limit. Of course, it either takes time or a large amount of hardware to perform the iterative decoding limiting the applications to low data rates where the processing can be done in software or to big budget projects where every last dB of performance is required.

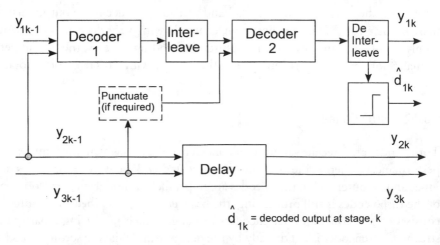

$\hat{d}_{1k}$ = decoded output at stage, k

Figure 6.53 Turbo code decoding module.

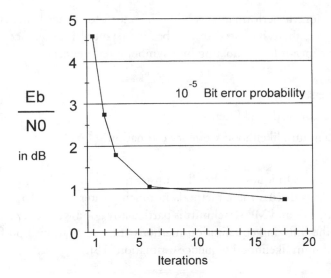

Figure 6.54 Half rate turbo code performance as a function of the number of iterations [50].

6.6.4 Trellis Code Modulation (TCM)

Frequently, error correcting coding is married with the modulation technique. The use of QPSK with 1/2 rate codes is a natural combination. Trellis code modulation [51] is a more exotic technique combining an error correcting code with QAM modulation. The basic TCM uses a 1/2 rate convolutional code to encode one bit of the m-bit symbol input to a QAM modulator. This produces an $m+1$ bit input to the symbol encoder. If a systematic code is used, the original m-bit word is input to the symbol encoder together with the parity bit. The convolutional encoder output is used to partition the QAM signal set into four sets. Signal points in the signal sets are separated by a distance of four (compared to two for the uncoded modulation), increasing the noise margin by 6 dB, if the sets can be identified. Adding the additional bit increases the signal energy by 3 dB so that a net increase in performance of 3 dB is obtained. It is possible to increase performance by 3 to 6 dB depending on the number of states. The price paid for the performance improvement is in the receiver complexity and susceptibility to demodulator phase errors.

6.7 Equalization

Intersymbol interference is a significant problem in many systems. In some cases, the PCM signal is deliberately filtered to control the signal spectrum and the degradation in performance is taken in return for other benefits. In most cases, however, there is a need to combat intersymbol interference. There are three generic means for combating intersymbol interference:

- Inverse filtering;
- Transverse digital filtering;
- Maximum likelihood sequence estimation.

Each of these techniques are briefly considered in the following sections. A number of texts discuss these methods in much greater detail [15,23–25]. The discussion by Lee and Messerschmitt is particularly good with an excellent comparison of the differing equalization methods and the relationship of equalization to maximum likelihood sequence estimation [15].

6.7.1 Filtering

The most straightforward approach uses a compensation filter which has the inverse response of the signal filtering transfer function. While this filter, if realizable, removes the overall intersymbol interference, it also enhances the noise at the decision point, degrading performance. The degree to which the noise is enhanced depends on the characteristics of the inverse filter and can be computed by

$$\frac{\sigma^2}{\sigma_0^2} = \frac{1}{\sigma_0^2} \int\limits_{-\infty}^{+\infty} N(f) \left| H_{eq}(f) \right|^2 df \qquad (6.125)$$

where

σ_0^2 = the mean square noise before equalization
$N(f)$ = the noise spectral density
$H_{eq}(f)$ = the equalizer response

If the response to be equalized has nulls, the equalizer is unrealizable and an approximation has very large gain at the null frequencies resulting in unacceptable noise enhancement. A more reasonable approach to equalization would be to attempt to balance the degradation due to the mean square intersymbol interference error with the mean square noise error. Furthermore, the intersymbol interference needs only to be compensated at the sampling times.

6.7.2 Transverse Digital Equalizers

The sampled output of a filter matched to the received signal is known to constitute a set of sufficient statistics [4] for the optimum detection of the data sequence. The noise samples are Gaussian but nonwhite. The noise samples can be whitened by passing them through a digital filter and the combination of the matched filter with the whitening filter is called a "sample-whitened matched filter." The equivalent discrete time model for the system incorporates the sample-whitened matched filter as shown in Figure 6.55.

The first type of digital equalizer (also called a transversal equalizer) is shown in Figure 6.56. The tap coefficients are adjusted to minimize the mean square error between the received sequence and the linear equalizer output. The tap coefficients can be shown to be the solution of

$$\sum_{i=0}^{L} c_i r_{i-k} = f_{k+d}$$

$$r_{i-k} = \sum_{0}^{L} f_j f^*_{i-k+j} + E\{n_k n_i\} \tag{6.126}$$

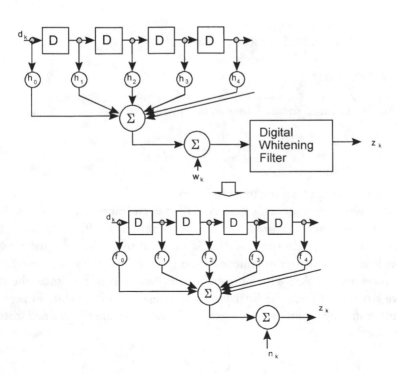

Figure 6.55 Discrete time model.

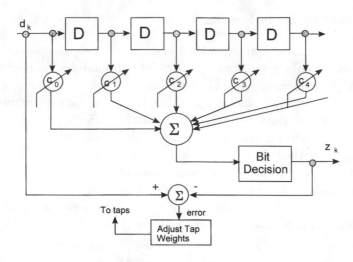

Figure 6.56 Linear transversal equalizer.

where

c_i are the tap weights
r_{i-k} = the sampled autocorrelation
f_k = the model tap weights
d = a delay

The minimum mean square error can be shown to be

$$\lambda^2{}_{LE} = 1 - \sum_0^L c_i f_{i+d} \qquad (6.127)$$

With no noise, the mean square error is zero.

The decision feedback equalizer (DFE) is an improvement on the linear equalizer as shown in Figure 6.57. This equalizer has feedforward and a feedback section. The feedforward section is identical to the linear equalizer while the feedback section tries to remove the portion of the intersymbol interference from previous samples. The tap gains can be adjusted to minimize the mean square error or to force the intersymbol interference components to zero. For the minimum mean square error method, the tap gains are determined from the solution of

$$\sum_0^{L_1} c_i r_{i-k} - \sum_1^{L_2} d_i f^*{}_{k+d-i} = f^*{}_{k+d} \qquad (6.128)$$

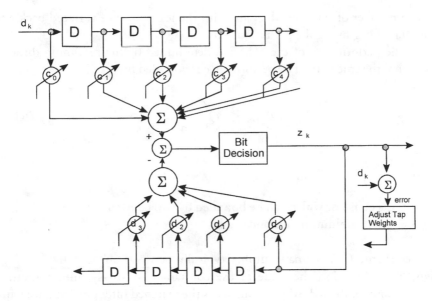

Figure 6.57 Decision feedback equalizer.

This reduces to the linear equalizer when the feedback terms are zero. The minimum mean square is more difficult to compute but it can be shown that the DFE error is less than the linear equalizer.

6.7.3 Maximum Likelihood Sequence Estimation (MLSE)

There is an alternative to the conventional equalizer which was suggested by the fact that the sample whitened matched filter output constitutes a set of sufficient statistics to estimate the entire data sequence. While this is now accepted as obvious, at one time the implications were not realized. The discrete time model looks exactly like the model of a convolutional encoder except that the tap weights have continuous real values instead of zero and one. Remembering that the idea of the Viterbi decoder is to estimate the states of the encoder, this suggests that the same approach can be used to estimate the data sequence. If the sequence of states are estimated, the symbols causing the state transitions are obtained. A system with L delay stages can be represented by a trellis diagram with 2^{L-1} states. With binary data, two paths enter each state and a real metric is associated with each path. As in a VA decoder for convolutional codes, at each symbol time the accumulated distance between the received sample and the two possible paths is computed and the branch with the least distance is labeled.

After a number of symbols, the path with the least distance is traced back and the data bit beginning the path is output.

The performance of the MLSE is determined by the minimum distance path from the true path. The performance is bounded by

$$P_{MLSE} \leq CQ\left(\frac{d_{min}}{2\sigma}\right) \tag{6.129}$$

where

C is a constant which can be bounded by two constants, $C_L \leq C \leq C_U$
d_{min} is the minimum distance path.

The constant, C, is a small number typically between one and five. At high signal-to-noise ratios, the value of the constant is relatively unimportant. A lower bound on the MLSE performance is the matched filter performance without intersymbol interference. It can be shown that in many cases the MLSE approaches, or equals, the matched filter performance. For channels with a response, $1 + \alpha D$, the performance is equal to the matched filter bound. This includes both duobinary and dicode signaling.

Finding the minimum distance path can be a problem for a channel with a large number of states. The VA can be used to find the minimum distance path by keeping track of the distance between the correct path and the competing paths. Although the MLSE offers optimal performance, there are a number of

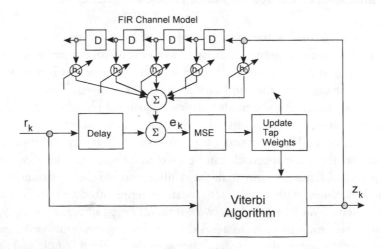

Figure 6.58 Maximum likelihood sequence estimator.

practical reasons for using DFE instead. First, the receiver complexity increases exponentially with the length of the channel impulse response and requires that the channel can be modeled as an FIR filter. Secondly, the VA requires a knowledge of the channel tap weights. If the tap weights are not known, they must be estimated. One means of estimating the channel weights is shown in Figure 6.58. The VA is started with an initial guess at the channel impulse response. The preliminary detected bit stream is fed into a linear transversal filter whose output is compared to the received data stream. The mean square error is used to update the transversal filter tap coefficients and the VA model coefficients. This receiver is a DFE with the VA as a detector. Thus, the MLSE is significantly more complex (and costly) than the DFE. With error correcting coding, the performance of the simpler technique can approach the optimum performance with much less complexity.

References

General

[1]   Panter, P. F., *Modulation, Noise and Spectral Analysis*, New York: McGraw-Hill Book Co., 1965.

[2]   Viterbi, A. J., *Principles of Coherent Communication*, New York: McGraw-Hill Book Co., 1966.

[3]   Schwartz, M., W. R. Bennett, and S. Stein, *Communication Systems and Techniques*, New York: McGraw-Hill Book Co., 1966.

[4]   Van Trees, H. L., *Detection, Estimation and Modulation Theory, Parts I, II and III*, New York: John Wiley and Sons, 1968.

[5]   Weber, C. I, *Elements of Detection and Signal Design*, New York: McGraw-Hill Book Co., 1968.

[6]   Helstrom, C. W., *Statistical Theory of Signal Detection*, Oxford: Pergamon Press, 1968.

[7]   Stiffler, J. J., *Theory of Synchronous Communications*, New Jersey: Prentice-Hall, 1971.

[8]   Clark, A. P., *Principles of Digital Data Transmission*, London: Pentech Press, 1976.

[9]   Melsa, J. L. and D. L. Cohn, *Decision and Estimation Theory*, New York: McGraw-Hill Book Co., 1978.

[10] Shanmugan, K. S., *Digital and Analog Communication Systems*, New York: John Wiley and Sons, 1979.

[11] Viterbi, A. J., and J.K. Omura, *Principles of Digital Communication and Coding*, New York: McGraw-Hill Book Co., 1979.

[12] Freeman, R. L., *Telecommunication System Engineering*, New York: John Wiley and Sons, 1980.

[13] Feher, K., *Digital Communications: Microwave Applications*, New Jersey: Prentice-Hall, 1981.

[14] Feher, K., *Digital Communications: Satellite/Earth Station Engineering*, New Jersey: Prentice-Hall, 1983.

[15] Lee, E. A. and D. G. Messerschmitt, *Digital Communication*, 2nd Ed, Boston: Kluwer Academic Publishers, 1994.

[16] Waggener, Bill, *Pulse Code Modulation Techniques*, New York: Van Nostrand Reinhold, 1995.

Error Correcting Coding References

[17] Forney, G. D., *Concatenated Codes*, Cambridge, MA: The MIT Press, 1966.

[18] Berlekamp, E. R., *Algebraic Coding Theory*, New York: McGraw-Hill Book Co., 1968.

[19] Hamming, R. W., *Coding and Information Theory*, New Jersey: Prentice-Hall, 1980.

[20] Peterson, W. W. and E. J. Weldon, *Error-correcting Codes,* 2nd Ed., Cambridge, MA: The MIT Press, 1984.

[21] Arazi, B., *A Commonsense Approach to the Theory of Error Correcting Codes*, Cambridge, MA: The MIT Press, 1988.

[22] Pretzel, O., *Error-Correcting Codes and Finite Fields*, Oxford, UK: Clarendon Press, 1992.

Equalization References

[23] Honig, M. L. and D. G. Messerschmitt, *Adaptive Filters: Structures, Algorithms, and Applications*, Boston: Kluwer Academic Publishers, 1984.

[24] Giodano, A. A. and F.M. Hsu, *Least Square Estimation with Applications to Digital Signal Processing*, New York: John Wiley and Sons, 1985.

[25] Alexander, S. T., *Adaptive Signal Processing: Theory and Applications*, New York: Springer-Verlag, 1986.

Other References

[26] Nyquist, H., "Certain Factors Affecting Telegraph Speed," *Bell System Technical Journal*, 31, 1924, pp. 324-346.

[27] Shannon, C.E., "A Mathematical Theory of Communication," *Bell System Technical Journal*, 27, July/October 1948.

[28] Forney, G. D., "Maximum-Likelihood Sequence Estimation of Digital Sequences in the Presence of Intersymbol Interference," *IEEE Transactions on Information Theory*, IT-18, May 1972, pp. 363-378.

[29] Messerschmitt, D.G., "A Geometric Theory of Intersymbol Interference, Part II: Performance of the Maximum Likelihood Detector,"*Bell System Technical Journal*, 52, November 1973.

[30]   Abramowitz, M., and Stegun, I., ed. *Handbook of Mathematical Functions*, National Bureau of Standards, AMS 55, June 1964.

[31]   Papoulis, A., *Probability, Random Variables, and Stochastic Processes*, New York: McGraw-Hill Book Co., 1965.

[32]   Pasupathy, S., "Minimum Shift Keying: A Spectrally Efficient Modulation," *IEEE Communications Magazine*, July 1979, pp. 14-22.

[33]   Austin, M., et al, "QPSK, Staggered QPSK, and MSK-A Comparative Evaluation," *IEEE Transactions on Communications*, COM-31, February 1983, pp. 171-182.

[34]   McLane, P., "The Viterbi Receiver for Correlative Encoded MSK Signals," *IEEE Transactions on Communications*, COM-31, February 1983, pp. 290-295.

[35]   Sundberg, C., "Continuous Phase Modulation," *IEEE Communications Magazine*, 24, April 1986, pp. 25-35.

[36]   Simmons, S.J. and Wittke, P.H., "Low Complexity Decoders for Constant Envelope Digital Modulations," *IEEE Transactions on Communications*, COM-31, December 1983, pp. 1273-1280.

[37]   Pizzi, S. and Wilson, S., "Convolutional Coding Combined with Continuous Phase Modulation," *IEEE Transactions on Communications*, COM-33, January 1985, pp. 20-29.

[38]   Galko, P. and Pasupathy, S., "Linear Receivers for Correlatively Coded MSK," *IEEE Transactions on Communications*, COM-33, April 1985, pp. 338-347.

[39]   Sevensson, A and Sundberg, C., "Serial MSK-Type Detection of Partial Response Continuous Phase Modulation," *IEEE Transactions on Communications*, COM-33, January 1985, pp. 44-52.

[40]   Simon, M., et al., *Spread Spectrum Communications, Volumes I, II and III*, Rockville, Maryland: Computer Science Press, 1985.

[41]   Dixon, R., *Spread Spectrum Systems, Second Edition*, New York: John Wiley and Sons, 1984.

[42]   Nicholson, D., *Spread Spectrum Signal Design*, Rockville, Maryland: Computer Science Press, 1988.

[43]   Davidson, F. and Sun, X., "Gaussian Approximation Versus Nearly Exact Performance Analysis of Optical Communication Systems with PPM Signaling and APD Receivers," *IEEE Transactions on Communications*, 11, November 1988, pp. 1185-1192.

[44]   Timor, U., and Linke, R., "A Comparison of Sensitivity Degradations for Optical Homodyne versus Direct Detection of On-Off Keyed Signals," *Journal of Lightwave Technology*, 11, November 1988, pp. 1782-1788.

[45]   Linke, R. and Gnauck, A., "High-Capacity Coherent Lightwave Systems," *Journal of Lightwave Technology*, 11, November 1988, pp. 1750-1769.

[46]   Rice, S.O., "Noise in FM Receivers," *Proceedings of Symposium of Time Series Analysis*, New York: John Wiley and Sons, 1963.

[47]   Schilling, D., Hoffman, E. and Nelson, E., "Error Rates for Digital Signals Demodulated by an FM Discriminator," *IEEE Transactions on Communication Technology*, April 1967.

[48]   Gradshteyn, I.S., and Ryzhik, I.M., *Table of Integrals, Series, and Products, Fifth Edition*, Boston: Academic Press, Inc., 1994, pp. 677 .

[49] Berlekamp, E., "The Technology of Error-Correcting Codes," *Proceedings of the IEEE*, 68, May 1980, pp. 564-593.

[50] Berrou, C. and Glavieux, "Near Optimum Error Correcting Coding and Decoding: Turbo-Codes," *IEEE Transactions on Communications*, 44, October 1996, pp. 1261-1271.

[51] Ungerbroeck, G., "Trellis-Coded Modulation with Redundant Signal Sets Part I: Introduction," IEEE Communication Magazine, 25, February 1987, pp. 5-11.

7

Synchronization

Although detection often gets top billing in discussions of PCM systems, synchronization is equally important. The matched filter must be sampled at the right time, the reference carrier in a PSK system must have the correct frequency and phase, the start of an error coded block must be known, and so on. All of these are manifestations of synchronization. From a systems standpoint, the importance of understanding synchronization cannot be overestimated. The time scale of the synchronization processes in the PCM system is illustrated in Figure 7.1. The process of achieving symbol and carrier synchronization is in the order of ten to hundreds of symbol periods. Code synchronization can take hundreds to thousands of symbol periods while format synchronization can take tens of thousands of symbol periods.

Three factors are of primary interest in the synchronization process: acquisition time, time to lose a lock, and synchronization threshold. Residual timing error is also of importance as a contributor to symbol error probability degradation. The discussion of synchronization will be separated into carrier, symbol, code, and format synchronization. Before considering each of these categories, the basic concepts of synchronization are discussed.

7.1 Synchronization Basics

Carrier synchronization is the problem of recovering a reference carrier with the same nominal frequency and phase as the received signal. Similarly, symbol synchronization is the problem of reconstructing a symbol rate timing reference.

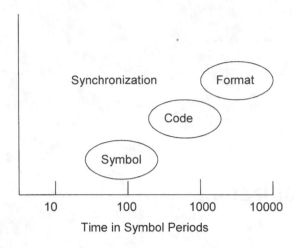

Figure 7.1 Time scale of synchronization processes.

Code and format synchronization require a reference identifying the beginning of a code block or a data frame. In the first two cases, the reference signal is a periodic function, either a sinusoidal or a pulse function. In a few cases, the received signal contains a reference component which may be noisy but can be used directly to extract a local reference. In most cases, the received signal must be processed to extract the local reference.

  Modulated carriers can be coherently demodulated if a local carrier reference with the same frequency and phase can be extracted from the received signal. The phase locked loop (PLL) is a component which can "lock" a local voltage controlled oscillator (VCO) to a periodic input such as a sinusoid. The lock condition implies that the reference VCO has the same frequency and phase apart from noise induced errors or systematic biases. The PLL is an integral part of most synchronization schemes and a review of the PLL provides a basis for considering the problems of carrier and symbol synchronization.

  The PLL has been extensively studied [1,2] and only the fundamental concepts are reviewed. The basic PLL is shown in Figure 7.2. The input signal will be assumed to be a cosine function

$$y(t) = A\cos(\omega_0 t + \theta_0) \tag{7.1}$$

The VCO reference carrier is

$$r(t) = \sin(\omega_0 t + \vartheta) \tag{7.2}$$

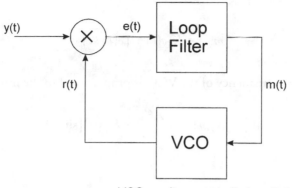

VCO = voltage controlled oscillator

Figure 7.2 Basic phase locked loop.

The phase difference between the input carrier and the reference is defined as

$$\phi = \vartheta - \theta_0 \qquad (7.3)$$

With a bit of trigonometry, the output of the multiplier (i.e., a balanced mixer) is

$$e(t) = \frac{A}{2}\sin\phi + \text{twice frequency terms} \qquad (7.4)$$

The twice frequency component is assumed to be removed by the linear loop filter. The loop filter is represented as a Heaviside differential operator.

$$m(t) = H_L(p)\{e(t)\} = \frac{A}{2}H_L(p)\{\sin\phi\} \qquad (7.5)$$

where

H_L is a rational polynomial in the differential operator, p.

The operator, p, is a shorthand notation for the differential operation and p^{-1} then stands for integration. If

$$H(p) = K\left(1 + \frac{1}{p}\right) \qquad (7.6)$$

The filter output would be

$$m(t) = K\left(e(t) + \int e(t)\, dt\right) \tag{7.7}$$

By definition, the frequency of the VCO is proportional to the input, $m(t)$.

$$\dot{\vartheta} = K_{VCO}m(t) = \frac{AK_{VCO}}{2} H_L(p)\{\sin\phi\} \tag{7.8}$$

where

K_{VCO} is the VCO sensitivity in radians per second per volt.

If the input carrier phase is assumed to be constant, the PLL is described by the nonlinear differential equation

$$\dot{\phi} = \frac{AK_{VCO}}{2} H_L(p)\{\sin\phi\} \tag{7.9}$$

This equation can be represented by the block diagram shown in Figure 7.3.

The multiplier is the phase detector and is ideally represented by a sinusoidal nonlinearity. If the PLL is near lock with a small phase error, the PLL can be modeled by the linear block diagram shown in Figure 7.4.

The transfer function between the input phase and the VCO phase is readily shown to be

$$\frac{\theta_{out}}{\theta_i} = \frac{K_d K_{VCO} H_L(s)}{s + K_d K_{VCO} H_L(s)} \tag{7.10}$$

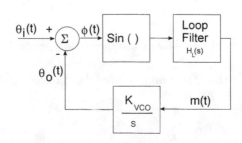

Figure 7.3 PLL equivalent block diagram.

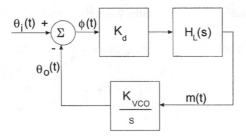

Figure 7.4 Small phase error model of a PLL.

Similarly, the phase error response is

$$\frac{\theta_e}{\theta_i} = \frac{s}{s + K_d K_{VCO} H_L(s)}$$ (7.11)

where

K_d is the equivalent phase detector gain

For a loop filter with a constant gain, the transfer function is a single pole low-pass response.

$$G_1(s) = \frac{\omega_n}{s + \omega_n}$$ (7.12)

where

$$\omega_n = K_d K_{VCO} H_0$$
$$H_0 = 1$$

This PLL is called a first order loop after the highest power of the denominator. The phase error response is

$$E_1(s) = \frac{s}{s + \omega_n}$$ (7.13)

The phase error due to a step change in phase is

$$e(t) = \Delta\theta e^{-\omega_n t}$$ (7.14)

The phase error decays exponentially with time and is 10% of the step value when $\omega_n t = 2.3$.

A more common PLL uses a loop filter with a characteristic response

$$H_L(s) = 1 + \frac{\alpha}{s} \tag{7.15}$$

The transfer function is

$$G_2(s) = \frac{K_d K_{VCO} s + K_d K_{VCO} \alpha}{s^2 + K_d K_{VCO} s + K_d K_{VCO} \alpha} \tag{7.16}$$

Historically, this transfer function has been expressed as

$$G_2(s) = \frac{2\zeta\omega_n s + \omega^2{}_n}{s^2 + 2\zeta\omega_n s + \omega^2{}_n} \tag{7.17}$$

where

ω_n is the loop "natural" frequency;
ζ is the "damping" factor.

This is a second order PLL with two poles and one zero. If the damping factor is reduced to a small value, the loop will have a tendency to oscillate at the natural frequency. The phase error response of the second order PLL is

$$E_2(s) = \frac{s^2}{s^2 + 2\zeta\omega_n s + \omega_n^2} \tag{7.18}$$

Second order PLLs are commonly used with a damping factor of 0.7. Higher order loops can be unstable and offer little advantage over the second order loop. The time required for a second order loop with $\zeta = 0.7$ to reduce the phase error to 10% of a step phase input is $\omega_n t = 3.6$.

The linear PLL analysis is limited to relatively small phase error changes. As input phase changes increase, the effective phase-detector gain decreases, increasing the response time of the loop. For large phase changes the loop skips cycles when trying to acquire the input signal. A point is reached at which the loop can no longer acquire the signal. The analysis of PLL acquisition is quite complex and relies on approximations and simulations to obtain results.

The acquisition process can be broken into two phases, frequency acquisition and phase acquisition. Frequency acquisition is the time required to reduce the input frequency error to the point where phase lock can be achieved without cycle slipping. Phase acquisition is the time required to reduce the phase error to near zero. Since frequency is the rate of change of phase, as the phase error changes, there is a corresponding change in instantaneous frequency.

The behavior of the PLL can be examined using a phase plane plot which plots frequency error versus phase error as illustrated in Figure 7.5. The trajectories of phase and frequency errors in the loop are plotted on this plane. The trajectory labeled "A" shows an initial phase error with no frequency error. As the loop tries to lock, the instantaneous frequency and phase change as shown with the trajectory eventually reaching the zero error state. Trajectory "B" shows a case in which an initial frequency error is acquired. The trajectory labeled "C" is interesting because it fails to lock within the limits of the plot. In this case, the PLL is skipping cycles to try to achieve lock. After one cycle, the loop has failed to reach lock but the frequency error has been reduced by δf. Perhaps this will reduce the frequency error sufficiently to allow the loop to capture the signal in the next cycle.

From the phase plane plot, the frequency acquisition time is defined as the time required to reduce the frequency error to the point at which the loop can acquire phase lock within one cycle. One might wonder what process allows the PLL to lock onto a signal with a large frequency error. A hand waving explanation of the acquisition process can be useful in understanding the PLL and its limitations. If the input signal is offset in frequency from the VCO, the output of the mixer has a beat frequency component equal to the difference between the input carrier frequency and the VCO frequency. This component will frequency modulate the VCO producing sideband frequencies about the VCO

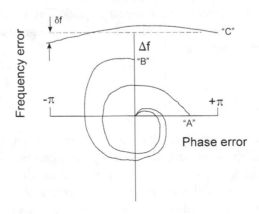

Figure 7.5 Phase plane.

center frequency. One of these components is at the input carrier frequency and the resulting DC component will bias the VCO frequency in the direction of input carrier. Eventually the PLL will lock onto the input carrier with an offset in the VCO frequency. In order for this process to work, the loop filter must pass the beat frequency component. If the amplitude of the beat frequency component is strongly attenuated, the VCO sideband at the input frequency will not be sufficiently strong to achieve frequency lock.

A number of individuals [1,4,5] have investigated the acquisition time for PLLs using approximations or simulations. Viterbi [1] used phase plane simulations to approximate the frequency acquisition time

$$T_{freq} \approx \frac{\Delta\omega^2}{2\zeta\omega_n^3} \tag{7.19}$$

The normalized frequency acquisition time for a loop with a damping factor of 0.7 is a function of the square of the frequency offset relative to the loop natural frequency.

$$f_n T_{freq} \approx \frac{1}{4\pi\zeta}\left(\frac{\Delta f}{f_n}\right)^2 \tag{7.20}$$

where

f_n = the loop natural frequency in hertz.

This is plotted in Figure 7.6 and can be used as a reasonable approximation for moderate frequency offsets. There is a frequency step offset below which the PLL will acquire the carrier without skipping cycles. This is called the "pull out" frequency. Simulation results for a second order loop suggest an approximation to the pull out frequency

$$\frac{\Delta f}{f_n} \approx 1.8(1+\zeta) \tag{7.21}$$

For a damping factor of 0.7, the loop can lock onto a frequency offset up to about three times the loop natural frequency. In a practical loop, the limit at which the loop can lock without cycle slipping is often less than twice the loop bandwidth. Although in theory, the PLL can lock with infinitely large frequency offsets, in practice, inherent noise and other factors limit the largest frequency offset to about 10 times the loop bandwidth.

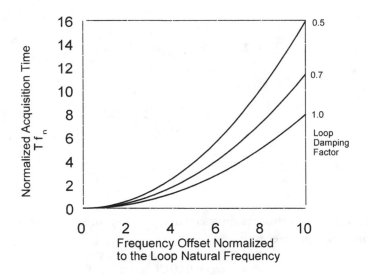

Figure 7.6 Normalized frequency acquisition time.

The effect of noise on the performance of the PLL can be estimated by assuming the noise is additive, Gaussian approximated as a bandpass process.

$$\tilde{n}(t) = \tilde{n}_I \cos\omega_0 t - \tilde{n}_Q \sin\omega_0 t \tag{7.22}$$

The output of the phase detector, assuming a small phase error, is approximately

$$e(t) \approx \frac{A}{2}\sin\phi + \frac{\tilde{n}_I}{2} \tag{7.23}$$

Higher order frequency terms are neglected leading to a model of the PLL with noise as shown in Figure 7.7. The mean square phase error due to the noise is

$$\sigma_\phi^2 = \int\limits_{-\infty}^{+\infty} \Phi(\omega)|G_L(\omega)|^2\, dw \tag{7.24}$$

where

G_L = the loop transfer function
Φ = the noise spectral density

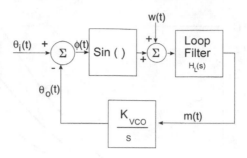

Figure 7.7 Noise model of PLL.

If the noise spectral density is approximately constant over the loop bandwidth, the mean square phase error is

$$\sigma_\phi^2 \approx \Phi(0)B_N \tag{7.25}$$

where

B_N = the equivalent noise bandwidth of the PLL

For a second order loop, the equivalent noise bandwidth is related to the loop natural frequency and damping factor by

$$\frac{B_N}{f_n} = \pi\left(\zeta + \frac{1}{4\zeta}\right) \tag{7.26}$$

The noise bandwidth is about three times the loop natural frequency for typical values of the damping factor. The mean square phase error for the linearized PLL is

$$\sigma_\phi^2 = \frac{1}{SNR_L} \tag{7.27}$$

where

SNR_L = the input signal-to-noise ratio in a bandwidth equal to the PLL noise bandwidth.

The performance of PCM systems is expressed in terms of the signal-to-noise ratio in a bandwidth equal to the bit rate so the loop bandwidth of the synchronizer should be much less than the bit rate to reduce the phase error variance to acceptable levels.

Viterbi [1] analyzes the more complex case of noise with the nonlinear differential equation model and is able to compute the probability density function of the phase error. An approximation to the phase error variance at low signal-to-noise ratios (0 to 6 dB) is

$$\sigma_\phi^2 \approx \frac{1}{SNR_L} + \frac{1}{SNR_L^2} \tag{7.28}$$

Given the probability density function and the phase variance, the probability that the PLL will lose lock due to noise can be computed. The probability density function can be obtained by solving the nonlinear differential equations known as the Fokker-Planck equations. Viterbi [1] solved these equations for a first order loop and obtained a density function of the form

$$p(\phi) = \frac{e^{\frac{\cos\phi}{\sigma^2}}}{2\pi I_0\left(\frac{1}{\sigma^2}\right)} \tag{7.29}$$

where

$$-\pi < \phi < \pi$$
σ^2 = the phase error variance
I_0 = the zeroth order modified Bessel function

This formidable expression can be approximated at high loop signal-to-noise ratios by a Gaussian distribution.

$$p(\phi) \approx \frac{1}{\sigma\sqrt{2\pi}} e^{-\frac{\phi^2}{2\sigma^2}} \tag{7.30}$$

The probability that the phase error will exceed some value, Θ, is

$$P(\phi \geq |\Theta|) = 2\int_\Theta^\infty p(\phi)\, d\phi = 2Q\left(\frac{\Theta}{\sigma}\right) \tag{7.31}$$

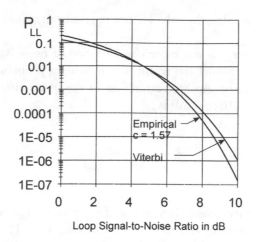

Figure 7.8 Probability of loss of lock.

Empirical data suggest a loss-lock probability of the form

$$P_{LL} \approx e^{-cSNR_L} \qquad (7.32)$$

where

　　　c is a constant in the range 1.5 to 3.0
　　　SNR_L = the loop signal-to-noise ratio

Assuming a threshold of 90 degrees, the probability of loss of lock as a function of loop signal-to-noise ratio is shown in Figure 7.8. With a loop width of 10% of the symbol rate, the PLL is capable of maintaining lock down to 0 dB or less.

　　　With this background on PLLs, it is time to apply the results to PCM synchronization applications.

7.2 Carrier Synchronization

Many PCM systems use some form of carrier modulation. With carrier modulation, optimum performance is obtained with coherent demodulation. The coherent demodulator requires a reference carrier which is frequency and phase coherent with the modulated carrier. The PLL is designed to phase lock a local carrier to a received carrier and is an obvious component of the coherent de-modulator. Carrier synchronization is used to extract a reference carrier from the received carrier in the presence of noise. The synchronization problem is straightforward if the received signal contains a spectral component at the carrier frequency or some multiple of the carrier frequency. In this case, the received signal is bandpass filtered about the unmodulated carrier component and a PLL is used to obtain the local reference. In general, the received signal does not contain an unmodulated component.

The focus of carrier synchronization is on amplitude modulation and phase modulation including continuous phase modulation, a form of FSK. Consider initially the amplitude modulation of a carrier signal by a digital data stream as shown in Figure 7.9. Double sideband, suppressed carrier (DSB-SC) is obtained by multiplying the carrier with a digital data stream. Coherent de-modulation is obtained by extracting a coherent reference and multiplying the received carrier by the reference. The receiver data filtering removes twice the frequency components and recovers the data. This type of modulation is commonly used with PAM and QAM signaling. The DSB-SC does not contain a purely periodic carrier component and the carrier synchronizer must process the signal to obtain a periodic component related to the carrier.

Conventional double sideband, amplitude modulation (DSB-AM) is generated by

$$y(t) = (1 + md(t))\cos\omega_c t \qquad (7.33)$$

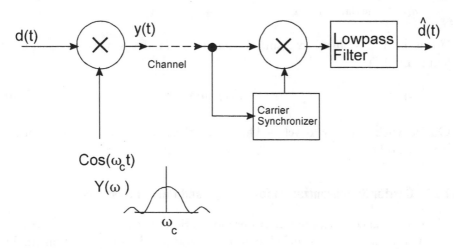

Figure 7.9 Double sideband, suppressed carrier modulation.

where

m = the modulation index
$d(t)$ = the digital modulation
$m\,d(t) < 1$

The constant term produces a discrete carrier component which can be used to extract the modulation. The desire to minimize bandwidth leads to vestigial sideband (VSB) and single sideband (SSB) systems. Both VSB and SSB can be represented by

$$y(t) = d_1 \cos\omega_c t + d_2 \sin\omega_c t \qquad (7.34)$$

These modulations are equivalent to the sum of an inphase and a quadrature phase DSB-SC signal.

The second generic modulation considered here is PSK, a special case of phase modulation in which the phase is shifted in discrete amounts in accordance with the data stream.

$$y(t) = A\cos(\omega_c t + md(t))$$
$$= A\cos(md)\cos\omega_c t - A\sin(md)\cos\omega_c t \qquad (7.35)$$

where

m = the phase modulation index
$d(t)$ = the data stream

When the data is binary (± 1) and the modulation index is $\pi/2$, PSK is equivalent to DSB-SC.

$$y(t) = d(t)\sin\omega_c t \qquad (7.36)$$

Carrier synchronization developed for one of these systems will also be applicable to the others.

7.2.1 Carrier Synchronization for Suppressed Carrier Systems

The performance of suppressed carrier modulations, such as PSK, is always better than a modulation which leaves an unmodulated carrier component. In some cases, residual carrier power or pilot tones may be used to simplify the receiver design at the expense of a performance degradation. Synchronization can be extracted from the suppressed carrier systems without having to sacrifice the loss of power. There are three common methods of carrier synchronization:

- Squaring loops;
- Costas loops;
- Decision directed loops.

7.2.1.1 Squaring Loop

The first obvious technique for extracting a carrier reference from a DSB-SC, or PSK signal is to square the signal using mixers as shown in Figure 7.10. In the absence of noise, squaring the signal results in

$$r(t) = y^2(t) = d^2(1 + \cos 2\omega_c t) \qquad (7.37)$$

Although the data is random, taking on ± 1, the squaring process removes the data modulation and creates a discrete spectral component at twice the input carrier frequency. The output of the squaring circuit is bandpass filtered and fed to a PLL. The PLL reference is divided by two to produce a reference at the carrier frequency.

7.2.1.2 Costas Loop

In an investigation of demodulation techniques for DSB-SC and SSB, Costas [6] proposed a special carrier synchronization loop which has become known as the Costas loop. The Costas loop is shown in Figure 7.11. The received signal is multiplied by inphase and quadrature phase outputs from a VCO. The mixer outputs are lowpass filtered to remove the twice frequency components. The resulting outputs are multiplied, filtered, and used to modulate the VCO.

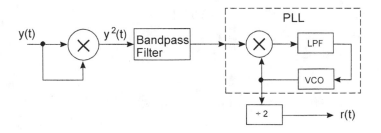

Figure 7.10 Squaring loop.

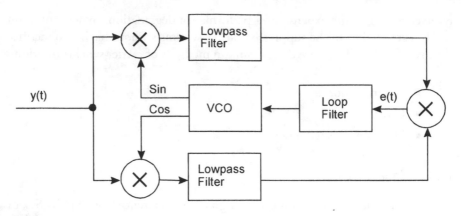

Figure 7.11 Costas loop.

The operation of the Costas loop can be explained as follows. Assume the input is the DSB-SC signal. Suppose the VCO has the correct frequency but a small phase error, $\delta\phi$. The quadrature reference carrier is

$$f_Q = \sin(\omega_c t + \delta\phi) = \cos\delta\phi \sin\omega_c t + \sin\delta\phi \cos\omega_c t \tag{7.38}$$

Similarly, the inphase reference carrier is

$$f_I = \cos(\omega_c t + \delta\phi) = \cos\delta\phi \cos\omega_c t - \sin\delta\phi \sin\omega_c t \tag{7.39}$$

If $\delta\phi$ is small,

$$\sin\delta\phi \approx \delta\phi$$

$$\cos\delta\phi \approx 1 \tag{7.40}$$

therefore,

$$f_Q \approx \sin\omega_c t + \delta\phi\cos\omega_c t$$

$$f_I \approx \cos\omega_c t - \delta\phi\sin\omega_c t \tag{7.41}$$

Remembering that the lowpass filters in each branch remove the second harmonic frequencies, the quadrature and inphase filter outputs are

$$g_Q \approx \frac{1}{2} K_1 d(t) \delta\phi$$

$$g_I \approx \frac{1}{2} K_2 d(t) \tag{7.42}$$

where

K_1 and K_2 are constants

The input to the VCO is the product of these outputs

$$e(t) \approx \frac{1}{4} K_1 K_2 d^2(t) \delta\phi \tag{7.43}$$

Variations due to the data stream are removed by using a narrow filter bandwidth relative to the data rate. The error signal to the VCO is proportional to the phase error and has the right sign to drive the loop toward lock. It is important to reiterate the requirement that the loop filter be narrow with respect to the data rate. A common mistake made by designers and users of carrier synchronizers is to ignore the data rate in selecting the carrier tracking loop bandwidth.

7.1.2.3 Decision Directed Loops

A third method of synchronization configuration is possible. Suppose the data modulation can be estimated. The modulation estimate can multiply the received carrier as shown in Figure 7.12 to remove the modulation.

$$r(t) = \hat{d}(t) d(t) \cos\omega_c t \tag{7.44}$$

If the estimate is perfect, an unmodulated carrier component is produced which can be used by a PLL to provide the reference carrier. Figure 7.13 is a block diagram of the decision directed loop. Once again, the operation of the loop is explained as follows. Assuming the VCO has the correct frequency but a small phase error, $\delta\phi$, the output of the inphase VCO component is

$$f_I = \cos(\omega_c t + \delta\phi) \approx \cos\omega_c t \tag{7.45}$$

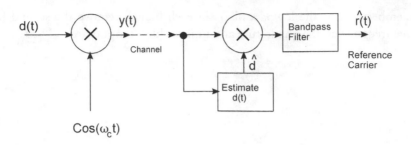

Figure 7.12 Removing carrier modulation.

The quaradture reference is

$$f_Q = \sin(\omega_c t + \delta\phi) \tag{7.46}$$

The inphase reference coherently demodulates the carrier and the matched filter provides an optimum estimate of the data. This estimate remodulates the quadrature carrier to be

$$f_m = \hat{d} \sin(\omega_c t + \delta\phi) \tag{7.47}$$

The data estimate is delayed one symbol period (the matched filter is sampled at the end of the symbol period) so a compensating delay of one symbol period is required in the quadrature path. The output of the quadrature mixer is

$$e(t) = d\,\hat{d}\,\sin\omega_c t\,\cos\omega_c t\,\cos\delta\phi + d\,\hat{d}\,\cos^2\omega_c t\,\sin\delta\phi \tag{7.48}$$

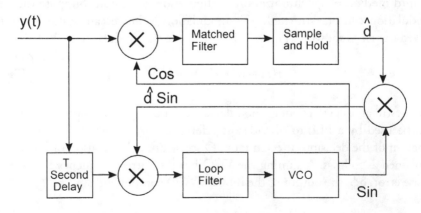

Figure 7.13 Decision directed loop.

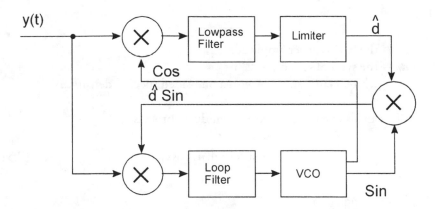

Figure 7.14 Approximation to decision directed loop.

The loop filter eliminates second harmonic terms and averages the product of the data and its estimate. The VCO input is then proportional to the phase error.

$$v(t) \approx K \sin \delta\phi \approx K \, \delta\phi \qquad (7.49)$$

The phase error term then drives the loop to lock.

The optimum decision directed loop requires symbol synchronization and a one symbol delay at the carrier frequency. The delay can be difficult to implement and must be tunable if multiple symbol rates are used. An approximation to the decision directed loop is shown in Figure 7.14. The data estimate is obtained using a lowpass filter and a limiter. The quadrature path delay is eliminated. The wider the lowpass bandwidth, the less the delay at the limiter and the poorer the data estimate. This suggests [7] that there is an optimum filter bandwidth which minimizes the loop mean square phase error. This is, in fact, true and the optimum lowpass filter bandwidth is in the order of twice the symbol rate. The suboptimum decision directed synchronization pays a performance penalty of about 3 dB relative to the optimum design.

7.2.2 Performance of PSK Synchronizers

The performance of each of the three types of suppressed carrier synchronizers is now considered. In each case the received signal is assumed to be of the form

$$y(t) = \sqrt{2} A \, d_k \, \sin \omega_c t + \bar{n} \qquad (7.50)$$

where

A = the RMS carrier amplitude
d_k = the symbol stream, = ± 1
n = white, Gaussian noise with a one-sided spectral density, N_0

The noise can also be represented as a bandpass process

$$\tilde{n} = \tilde{n}_1 \sin \omega_c t + \tilde{n}_2 \cos \omega_c t \qquad (7.51)$$

7.2.2.1 Squaring Loop

The output of the squaring circuit at frequencies near $2\omega_c$ is

$$y^2 \approx -\frac{\rho^2}{2} \cos 2\omega_c t + \rho \tilde{n}_2 \sin 2\omega_c t + \frac{\tilde{n}_2^2}{2} \cos 2\omega_c t \qquad (7.52)$$

where

$$\rho = \sqrt{2} A d_k + \tilde{n}_1$$

The VCO output is

$$f_{VCO} = \sin(2\omega_c t + 2\varphi) \qquad (7.53)$$

The error function at the mixer output is

$$e(t) = -\frac{A^2}{2} \sin 2\varphi + \tilde{N}_1 \sin 2\varphi + \tilde{N}_2 \cos 2\varphi \qquad (7.54)$$

where

N_1 and N_2 are functions of n_1, n_2 and d_k

As with the PLL, the loop filter output can be expressed in Heaviside notation

$$h(t) = H(p) \left\{ \frac{A^2}{2} \sin 2\varphi + \tilde{N}_1 \sin 2\varphi + \tilde{N}_2 \cos 2\varphi \right\} \qquad (7.55)$$

Finally, the squaring loop can be described by the nonlinear differential equation

$$\dot{\varphi} + K_{VCO}H(p)\left\{\frac{A^2}{2}\sin 2\varphi + \tilde{w}\right\} \tag{7.56}$$

where

$w =$ the equivalent noise

The differential equation suggests the equivalent block diagram shown in Figure 7.15. This is identical to a conventional PLL except for the phase detector gain and the more complex noise term. The acquisition time and lock-in performance is comparable to a conventional PLL with the same loop bandwidth.

Analysis of the noise performance is more complex but the high loop signal-to-noise ratio mean square phase error is identical to a PLL

$$\sigma^2 = \frac{1}{SNR_L} \text{ for high } SNR \tag{7.57}$$

For low SNR_L, the mean square phase error is approximately

$$\sigma^2 \approx \frac{3}{4SNR_L^2} \tag{7.58}$$

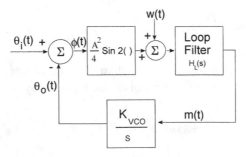

Figure 7.15 Equivalent block diagram of squaring loop.

Thus the mean square phase error as a function of SNR_L is approximately

$$\sigma_{SL}^2 \approx \frac{1}{SNR_L} + \frac{0.75}{SNR_L^2} \tag{7.59}$$

7.2.2.2 Costas Loop

The same analysis can be used for the Costas loop. Rather than repeat the cumbersome trigonometry, the Costas loop can be described by the identical non-linear differential equation (7.56) obtained with the squaring loop. Thus the Costas loop has the same theoretical performance as the squaring loop! It can be shown that the optimum lowpass filters for the Costas loop are matched filters although the performance is not strongly dependent on the actual filter response.

7.2.2.3 Decision Directed Loop

Neglecting second harmonic frequency terms, the output of the loop filter in the decision directed loop is

$$h(t) \approx H(p)\left\{ \frac{\sqrt{2}}{2} A d_k \hat{d}_k \sin\varphi + \frac{\hat{d}\tilde{n}_1}{2}\cos\varphi + \frac{\hat{d}\tilde{n}_2}{2}\sin\varphi \right\} \tag{7.60}$$

The equivalent differential equation becomes

$$\dot{\varphi} + K_{VCO}H(p)\left\{ \frac{\sqrt{2}A}{2} d_k \hat{d}_k \sin\varphi + \bar{w}_1 \right\} = 0 \tag{7.61}$$

where

$$\bar{w}_1 = \frac{\hat{d}\tilde{n}_1}{2}\cos\varphi + \frac{\hat{d}\tilde{n}_2}{2}\sin\varphi$$

If the loop bandwidth is much smaller than the symbol rate, the product of the symbol and the estimate of the symbol can be replaced by its expected value.

$$E\left\{\hat{d}\,d\right\} = (+1)prob\left(\hat{d} = d\right) + (-1)prob\left(\hat{d} \neq d\right)$$

$$= 1 - 2p \tag{7.62}$$

where

  p = probability of a symbol error

With this substitution, an equivalent block diagram for the decision directed loop is shown in Figure 7.16. Once again, the block diagram reduces to an equivalent PLL. The mean square phase error is

$$\sigma^2_{DDL} = \frac{1}{SNR_L(1-2p)^2} \qquad (7.63)$$

For large signal-to-noise ratios, the mean square phase error approaches the other types of synchronizers. At lower signal-to-noise ratios, the effects of symbol errors must be considered.

  With the suboptimum decision-directed loop using a limiter for detection, the mean square phase error is approximately

$$\sigma^2_{SDDL} \approx \frac{1}{SNR_L(1-2p)^2 \, R^2(\Delta t)} \qquad (7.64)$$

where

  $R(\Delta t)$ = the data autocorrelation function evaluated at the delay time

If a matched filter followed by a limiter is used as a detector, $R(\Delta t) = 0.5$, and the suboptimum loop has about twice the phase error at large SNR_L as compared to the optimum loop.

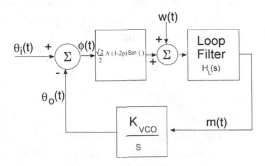

Figure 7.16 Equivalent block diagram for decision directed loop.

7.2.2.4 Maximum Likelihood Synchronizer

A theoretical approach can be taken to the design of the carrier synchronizer. Estimation theory can be applied to the problem of estimating the carrier phase in the presence of noise. In general, the problem is one of estimating the phase, ϕ, from the received signal, $z(t)$.

$$z(t) = \sqrt{2}A\cos\left(\omega_c t + \theta(t) + \phi\right) + \tilde{n}(t) \tag{7.65}$$

where

$\theta(t)$ = the phase modulation, assumed random
$n(t)$ = white, Gaussian noise
ϕ　 = the unknown phase, to be estimated

One method which can be used is the maximum likelihood estimator (MLE) which maximizes the posteriori conditional probability, $p(\phi z)$. From Bayes rule

$$p(\phi|z) = \frac{p(z|\phi)p(\phi)}{p(z)} \tag{7.66}$$

Since $p(z)$ is independent of ϕ, maximizing $p(\phi|z)$ is equivalent to maximizing

$$L = p(z|\phi)p(\phi) \tag{7.67}$$

The MLE assumes no prior knowledge of the probability density of ϕ so it maximizes $p(z|\phi)$, or any monotonic function of $p(z|\phi)$. The natural logarithm of $p(z|\phi)$ is more convenient to maximize.

$$\Lambda = \ln p(z|\phi) \tag{7.68}$$

The received signal is a function not only of ϕ but also the modulation, θ. The modulation effects will be eliminated by averaging $p(z|\phi,\theta)$ over θ. The development of the MLE is very messy and only a "hand waving" outline of the development is presented.

The conditional probability, $p(z|\phi,\theta)$ is obtained by sampling the signal, computing the joint density of the samples, and letting the limits of the sampled case approach the continuous case.

$$p(z|\phi,\theta) = C_1 e^{-\frac{1}{2N_0}\int(z(t)-s(t))^2 dt}$$

(7.69)

where

$s(t) = $ the signal portion of the signal

This can be simplified to

$$p(z|\phi,\theta) = C_2 e^{\frac{1}{N_0}\int z(t)s(t)dt}$$

(7.70)

The conditional probability, $p(z|\phi)$, is then found by taking the expected value with respect to θ.

$$p(z|\phi) = E_\theta\{p(z|\phi,\theta)\}$$

(7.71)

For the case of PSK, the modulation takes on two values, $+\pi/2$ and $-\pi/2$ with equal probability. The expectation can be performed resulting in

$$p(z|\phi) = C_3\left[e^{\frac{b(\phi)}{N_0}} + e^{-\frac{b(\phi)}{N_0}}\right] = C_3 \cosh\left(\frac{b(\phi)}{N_0}\right)$$

(7.72)

where

$$b(\phi) = \int_0^T z(t)\sqrt{2}\,A\sin(\omega_c t + \phi)dt$$

(7.73)

We wish to choose ϕ such that the log likelihood ratio is maximized.

$$\Lambda = \ln\left(C_3 \cosh\left(\frac{b(\phi)}{N_0}\right)\right)$$

(7.74)

Taking the derivative of Λ with respect to ϕ.

$$\frac{\partial \Lambda}{\partial \phi} = C_4 \tanh \frac{b(\phi)}{N_0} \frac{\partial b}{\partial \phi} \tag{7.75}$$

But

$$\frac{\partial b}{\partial \phi} = \int_0^T z(t)\sqrt{2}A\cos(\omega_c t + \phi)dt \tag{7.76}$$

To maximize the log likelihood ratio, (7.75) should be forced to zero by varying ϕ. This can be done with the circuit shown in Figure 7.17. The nonlinear function, *tanh(x)*, can be approximated by a limiter at high signal-to-noise levels and by a linear function at low levels. In the first case, the MLE is a decision directed loop while in the second case it is a Costas loop.

7.2.2.5 Comparision of Performance

The performance of the carrier synchronizers can be compared on the basis of mean square phase error as a function of loop signal-to-noise ratio. At high signal-to-noise ratios the mean square error should approach that of a PLL with an unmodulated carrier. The MLE synchronizer performance, which can be

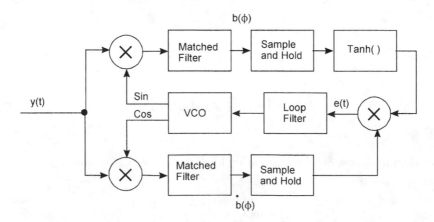

Figure 7.17 Maximum likelihood estimator for carrier synchronization.

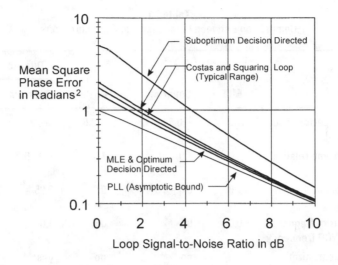

Figure 7.18 Comparison of carrier synchronizer performance.

bounded using the Cramer-Rao bound, is a lower bound on performance. Figure 7.18 compares the different synchronizer implementations.

The Costas and squaring loops have been shown, theoretically, to have identical performance. Two curves are shown for the Costas/squaring loop. The bound nearest the MLE bound assumes the use of ideal matched filters while the second curve assumes an ideal lowpass filter. When the optimum decision directed synchronizer is implemented with matched filter detection and perfect delay compensation, the performance is very near the MLE bound. When the limiter/lowpass filter decision directed synchronizer is implemented, the performance is degraded several dB as shown.

Although it is subjective, the relative implementation complexity has been estimated based on the number of components required and the synchronizers as compared in Table 7.1.

7.2.2.6 Extension to Multilevel Signaling

The three types of demodulators can be extended to multilevel signals. The performance of the decision directed, Costas, and squaring loops for QPSK have the same relative performance as for binary PSK. The squaring loop is the least complex to implement with the MLE, the most complex. Generally, the Costas synchronizer is the preferred implementation since one arm of the synchronizer provides the optimum demodulator for the modulated carrier.

Table 7.1
Comparison Between Components and Synchronizers

| Parameter | MLE | Squaring | Costas | Decision Directed | |
				Optimum	Limiter
Phase error performance relative to MLE in dB	1.0	−0.5 to −2	−0.5	−0.2	−1.8
Acquisition time relative to MLE	1.0	1.0	1.0	~1.15	~1.90
Implementation complexity	1.0	0.6 – 0.8	0.9	0.95	0.7
Highest carrier frequency relative to VCO frequency	0.5	0.5	0.5	1.0	1.0
Symbol synchronization required	yes	no	no	yes	no

7.3 Symbol Synchronization

If the PCM signal is a baseband signal, the symbol timing must be extracted to provide optimum detection and to be able to clock the serial data stream. The clock extraction is similar to the problem of carrier synchronization. Symbol synchronization differs in one major respect from carrier synchronization in the random nature of the signal. The received signal is of the form

$$z(t) = \sum \bar{d}\, p(t - kT) + \tilde{n}(t) \qquad (7.77)$$

where

d = the symbol values, ± 1
n = the noise

The symbols are assumed to be random so that there is a finite probability of a number of successive symbols being the same polarity. Thus, there are periods with no symbol transitions making it more difficult to extract a clock from the received signal.

The approaches to designing symbol synchronizers have either been based on ad hoc (intuitive) methods or derived from theoretical considerations. The intuitive designs have generally converged on the approximations to theoretical concepts. The intuitive approach tries to extract timing pulses from the data transitions. A PLL uses the transition pulses to phase lock a reference clock to the symbol rate. Figure 7.19 shows an example of a symbol synchronizer for a NRZ-L PCM signal. The derivative of the matched filter is used to identify symbol transitions. When there are no transitions, the derivative is zero. If the rising edge of the derivative is used to produce a 1/2 symbol pulse and the falling edge also produces such a pulse, the sum of the two pulse trains is periodic at the symbol rate. A PLL can lock on the derived pulse train even though there are periods with no transitions. A number of transition detectors have been proposed and implemented in commercially available symbol synchronizers.

The use of nonrectangular symbol waveforms, such as raised cosine and partial response, can simplify the transition detector and can use a squaring type of synchronizer similar to the carrier synchronizer to extract a clock reference. Some types of signals are designed to insure a data transition each symbol time.

A theoretical approach can be taken to symbol synchronization using the same estimation techniques discussed earlier. The results for symbol synchronization using MLE bear a close resemblance to the carrier results. The log likelihood ratio for symbol synchronization with binary symbols is

$$\Lambda(\tau) = C \sum \ln\left(\cosh\left[\frac{d}{N_0} y(\tau)\right]\right) \tag{7.78}$$

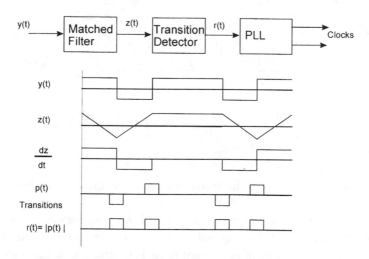

Figure 7.19 Symbol synchronizer for NRZ-L PCM.

where

$$y(\tau) = \int_0^T z(t)h(t - nT - \tau)dt$$

$y(\tau)$ = the sampled output of the matched filter
d = symbol amplitude
N_0 = the noise spectral density
C = a constant

The timing parameter is chosen such that the log likelihood is a maximum, or the derivative equals zero.

$$\frac{\partial \Lambda(\tau)}{\partial \tau} = C_1 \sum \left[\frac{d}{N_0} \dot{y}(\tau) \tanh\left(\frac{d}{N_0} y(\tau) \right) \right] \qquad (7.79)$$

Choosing τ to force this expression to zero suggests a feedback system as shown in Figure 7.20. The synchronizer looks very much like the optimum carrier synchronizer.

If the signal-to-noise ratio is low, the *tanh(x)* function is approximately linear and the loop is similar to a Costas loop. For high signal-to-noise ratio, the *tanh(x)* function is approaches a limiter and the synchronizer is a decision directed loop. The derivative of the log likelihood ratio can be approximated by finite differences.

$$\frac{\partial \Lambda(\tau)}{\partial \tau} \approx \frac{\Lambda\left(\tau + \frac{\Delta}{2} \right) - \Lambda\left(\tau - \frac{\Delta}{2} \right)}{\Delta}$$

$$\frac{\partial \Lambda(\tau)}{\partial \tau} \approx C \sum \frac{2d}{N_0} \left(\left| y\left(\tau + \frac{\Delta}{2} \right) \right| - \left| y\left(\tau - \frac{\Delta}{2} \right) \right| \right) \qquad (7.80)$$

This approximation leads to the "early-late" gate synchronizer as shown in Figure 7.21.

All of these synchronizers can be reduced to an equivalent block diagram as shown in Figure 7.22. The equivalent phase detector output is multiplied by the transition density and an equivalent noise is summed with the phase detector output.

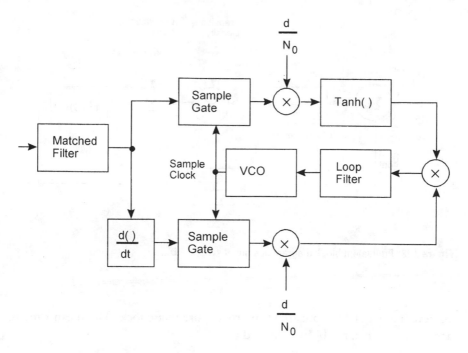

Figure 7.20 MLE symbol synchronizer.

7.3.1 Symbol Synchronizer Performance

The performance of the symbol synchronizer can be estimated using the PLL model. The acquisition time can be estimated using empirical and approximate theoretical results. The total time to acquire lock is estimated as the time to

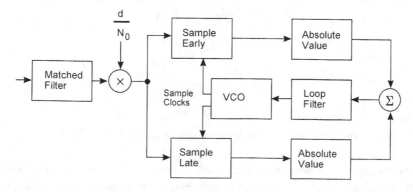

Figure 7.21 Early-late gate symbol synchronizer.

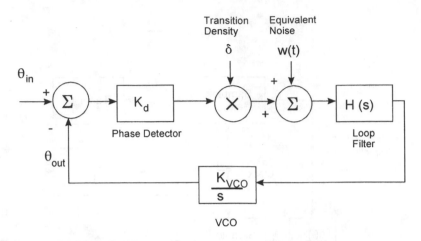

Figure 7.22 Equivalent block diagram of symbol synchronizer.

achieve frequency lock plus the time to acquire phase lock. The mean time to acquire phase lock has been estimated as

$$\overline{T}_{PL} \approx \frac{1.3}{B_N} \tag{7.81}$$

where

   B_N = the loop noise bandwidth

This time can be normalized as follows

$$\overline{T}_{PL}f_s = N_{PL} \approx \frac{0.41}{\left(\dfrac{f_n}{f_s} \right)} \tag{7.82}$$

where

   N_{PL} = the time to phase lock expressed in symbol periods.

The denominator is normally called the loop bandwidth and is usually normalized to the loop natural frequency rather than the loop noise bandwidth. A 1% loop will take about 41 symbol periods to achieve phase lock.

The time to achieve frequency lock is estimated to be

$$\overline{T}_{FL} \approx 4\frac{\Delta f^2}{B_N^3} \qquad (7.83)$$

where

Δf = the frequency offset

This expression can also be normalized.

$$\overline{T}_{FL}f_s = N_{FL} \approx 0.13\frac{\left(\dfrac{\Delta f}{f_n}\right)^2}{\left(\dfrac{f_n}{f_s}\right)} \qquad (7.84)$$

The total acquisition time is approximately

$$N_{ACQ} \approx \frac{0.41\left(1+0.32\left(\dfrac{\Delta f}{f_n}\right)^2\right)}{\left(\dfrac{f_n}{f_s}\right)} \qquad (7.85)$$

The acquisition time is also a function of transition density which is not apparent in these equations. The loop natural frequency and damping factor is designed for an average transition density, typically 0.5. If the average transition density varies from the design value, the acquisition time will vary accordingly. The lower the density, the longer the loop acquisition time will be. The loop natural frequency and the damping factor both vary as the square root of the transition density. Thus the acquisition time is expected to vary inversely with the square root of the transition density.

The total acquisition time in symbol periods is shown in Figure 7.23 as a function of frequency offset and loop width. The acquisition time estimates are consistent with measured data with frequency offsets less than the loop width but tend to be optimistic for large frequency offsets.

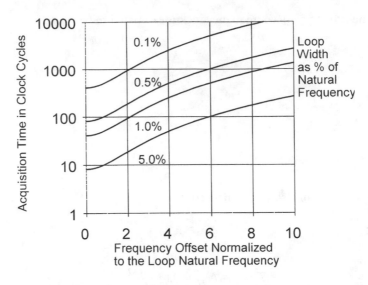

Figure 7.23　Symbol synchronizer acquisition time.

The mean square phase error can be estimated using the linearized model of the synchronizer. As with the carrier synchronizers, the mean square phase error is inversely proportional to the loop signal-to-noise ratio. With symbol synchronizers, the mean square timing error is normalized to the symbol period. An approximation to the mean square timing error of the MLE synchronizer is shown in Figure 7.24. The lower bound assumes 100% transition density. Decreasing transition density degrades loop signal-to-noise ratio and increases mean square timing error. The synchronization threshold is increased by about 3 dB each time the transition density is halved.

The importance of maintaining a high average transition density is apparent. The use of pseudo random scrambling provides an average density of 50% although long runs without a transition can occur. Often group coding is used to insure a minimum transition density. Error correcting coding also randomizes the symbol stream and maintains an average density of about 50%.

7.4　Code Synchronization

Once carrier and symbol synchronization are achieved, the serial data stream is stable with some residual bit error probability. If the data stream is coded, the decoder must synchronize the block (block codes) or the node (convolutional codes) before the error correction can occur. Block codes are fixed length blocks

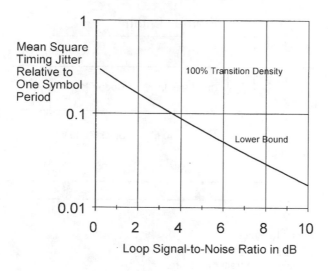

Figure 7.24 Symbol synchronizer timing error.

and the system must identify the starting bit of a block. Convolutional codes interleave bits from the code generators and the start of one of these bits must be identified. Sorting out the convolutional code bits is called node synchronization.

7.4.1   Block Codes

Synchronization of block codes can be accomplished by several methods. In many cases, the block code words are embedded in a frame structure as illustrated in Figure 7.25, with frame synchronization patterns inserted every N code words. The frame synchronization word (or marker) is chosen to be a non valid code and a frame synchronizer searches the input stream for the frame synchronization word. The synchronizer uses a sliding correlator which computes the distance between a group of symbols and the known pattern. The time of occurrence of the pattern with the minimum distance is taken as the most probable synchronization word location and the synchronizer may accept this location or may look in successive frames to see if it matches in the same location. In the best scenario, frame synchronization is obtained in one frame time. For a block code length of L symbols and a frame of N words, it takes a minimum of NL symbols to obtain code synchronization. Whereas symbol synchronization may take hundreds of symbols to achieve synchronization, the block code synchronization may take thousands of symbols.

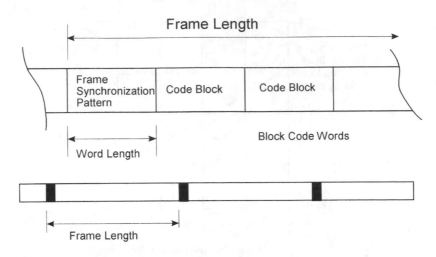

Figure 7.25 Block codes in a frame structure.

An alternative method of code synchronization keeps track of the code syndrome each symbol time along with an index to mark the relative symbol time. After L symbols, the symbol time with the smallest syndrome value is taken as a trial location for the end of a code block. If there are no symbol errors, the correct location will have a zero syndrome value. This process can be integrated over a number of code words using a technique as shown in Figure 7.26. A memory with L locations is used to hold an accumulated syndrome value for each of the possible block boundaries. After a number of blocks, the integrated

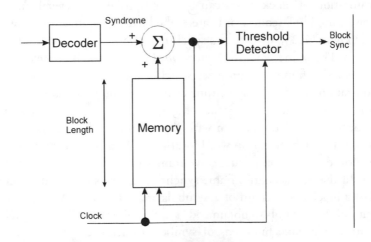

Figure 7.26 An integrating correlator word synchronizer.

value in the correct block location will be small while the incorrect locations will increase. A threshold can be set with the correct location corresponding to the memory location with the least value. This method eliminates the need for a special frame synchronization marker and can acquire code synchronization in less than a frame time.

7.4.2 Convolutional Codes

In a 1/2 rate convolutional code, alternate bits represent the output of the two code generators. Code synchronization (also called node synchronization) requires the receiver to identify which bits come from which code generator. Several methods have been proposed to acquire node synchronization. One method periodically inserts a frame synchronization marker into the data stream. If the marker is longer than the code constraint length, a fixed pattern occurs in the coded stream and the frame synchronization method discussed for block codes can be used.

A second method is similar to the syndrome detection technique discussed for block codes. Taking a 1/2 rate systematic code as an example, the syndrome can be computed each symbol time for each of the two possible symbol times. These two syndromes are integrated and compared. Eventually, the integrated syndrome for the correct node location will be less than the incorrect location.

The third method uses the state metrics in the Viterbi decoder to determine the correct node. When the decoder is in synchronization, there is always a correct path through the trellis diagram with one path (the most probable one) having the smallest accumulated distance. The path metric in correct node synchronization increases only because of channel noise. If the decoder has the wrong node synchronization, the path metrics will increase not only due to noise but also to the incorrect data sequence. By examining the path metrics, the correct node synchronization can be obtained. An analysis of the state metric method [8] shows that the detection threshold is the same for all convolutional codes with the same rate and constraint length. Node synchronization was obtained for some specific 1/2 rate codes in less than 1500 symbol times at a 2 dB signal-to-noise ratio.

7.5 Format Synchronization

The data in PCM applications has a word structure which may be simple or complex. Data words are typically organized into frames with distinct frame synchronization markers. The format may be relatively simple like the organization of words in a T-1 telecommunications frame or as complex as a telemetry frame with super commutation and subcommutation. These formats are illustrated in Figure 7.27. To extract the data from the format requires format synchronization [9] which identifies word boundaries. The problem of obtaining block code synchronization is one aspect of format synchronization.

Format synchronization is obtained by searching the incoming symbol stream for known synchronization patterns. Once a tentative pattern location is found, it is verified by looking for the pattern in successive frames. In complex formats it is necessary to locate both main frames and multiplexed subframes. The synchronization process can be further complicated by the use of variable length frames or by the use of asynchronous packets.

The performance of the format synchronizer is, obviously, format dependent. For simple formats with an embedded frame synchronization marker, the time to acquire synchronization depends on the length of the synchronization marker, the frame length, the bit error probability and the synchronization strategy. If the synchronizer requires M check frames before declaring lock, the frame synchronizer will acquire lock in about $1 + M$ frames for bit error probabilities less than 10^{-1}. For higher error rates, the acquisition time increases rapidly. Reference [9] provides detailed analyses of some typical telemetry formats.

The time to lose lock also depends on the format structure, synchronization strategy and the bit error probability. The typical lock strategy requires sev-

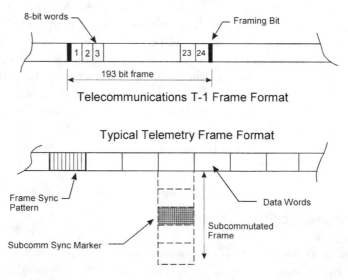

Figure 7.27 Typical PCM formats.

Table 7.2
Mean Time to Lose Lock in Frames and Bit Error

Bit Error Probability	Mean Time to Lose Lock in Frames			
	Consecutive Frames in Error			
	2	3	4	5
0.1	113	1112	1.1×10^4	1.1×10^5
0.01	$\sim 10^4$	$\sim 10^6$	$\sim 10^8$	$\sim 10^{10}$

eral consecutive framing errors before declaring a loss of frame synchronization. The mean time to lose lock expressed in frames can be approximated by

$$\overline{T}_{LL} \approx L + \frac{1}{(1-p)p^L} \qquad (7.86)$$

where

p = the bit error probability
L = the number of consecutive frames in error

For error probabilities less than 10^{-2} and more than three consecutive frame errors, the mean time to lose lock is more than 10^6 frames. The mean time to lose lock is summarized in Table 7.2 for several values of consecutive frames in error and bit error probability.

Format synchronization acquisition times are in the thousands of symbol periods for many applications. The format synchronization performance is only of concern for bit error probabilities greater than 10^{-2}. At these error levels the entire PCM system begins to rapidly deteriorate. PCM systems are truly digital, they often either operate without error or don't operate at all. The difference between error-free operation and complete failure is often only a few dB in signal-to-noise ratio.

7.6 Equalizer Training

Systems with serious intersymbol interference require equalization which must generally adapt to unknown or changing channel conditions. The system may not be considered stable in performance until the equalizer has converged to a set of tap coefficients close to the final values. The tap gains are computed according to

$$c_{k+1} = c_k + \Delta e_k \qquad (7.87)$$

where

c_k is the tap gain at time k
Δ = the incremental gain factor
e_k = the error signal

The incremental gain factor is typically small in order for the tap gain adaptation to be stable. If the error signal is normalized to a maximum value of one, the time in symbol periods required to reduce a tap gain difference of δc to zero is in the order of

$$T_c \approx \frac{\delta c}{\Delta} \qquad (7.88)$$

Depending on the severity of the intersymbol interference, the time for the equalizer to converge can range from tens to thousands of symbol times. If the channel changes slowly, once initial convergence has been obtained, subsequent changes are much faster. Thus, the time for equalization adaptation is typically of the same order as symbol synchronization and less than the time for format synchronization.

References

[1]   Viterbi, A., *Principles of Coherent Communication*, New York: McGraw-Hill Book Co., 1966

[2   Gardner, F., *Phaselock Techniques*, New York: John Wiley and Sons, 1966

[3]   Gruen, W., "Theory of AFC Synchronization," *Proceedings of the I.R.E.*, 38, February 1953, pp. 603-633.

[4]   Rey, T., "Automatic Phase Control: Theory and Practice," *Proceedings of the I.R.E.*, 48, October 1960, pp. 1760-1771.

[5]   Ibid, "Corrections to Automatic Phase Control: Theory and Practice," *Proceedings of the I.R.E.*, March 1961, p. 590.

[6]   Costas, J. P., "Synchronous Communications," *Proceedings of the I.R.E.*, 44, December 1956, pp. 1713-1718.

[7]   Waggener, W. N., "Optimum Design of PSK Demodulators," *Proceedings of the 8th International Aerospace Symposium*, Cranfield, U.K., March 1975.

[8]   Cheng, U., "Node Synchronization of Viterbi Decoders Using State Metrics," *TDA Progress Report 42-94*, Jet Propulsion Laboratories, April-June 1986.

[9]   Waggener, B., *Pulse Code Modulation Techniques*, New York: Van Nostrand Reinhold, 1995, pp. 313-353.

About the Author

William (Bill) Waggener has over 35 years of experience in communication, electro-optical, and recording systems. Mr. Waggener retired from Loral Data Systems in 1995 as Vice President of Engineering and formed his own research and development consulting confederacy, Chestnut Mountain Group (CMG). Most recently, he has been working on telecommunications and surveillance applications for high altitude, long endurance (HALE) vehicles in conjunction with B&R Designs, Inc., in Sarasota, Florida.

As a result of his work, Mr. Waggener has received 10 U.S. patents, published over 35 technical papers, and is the author of the recently published book, *Pulse Code Modulation Techniques*, by Van Nostrand Reinhold, published in 1995. He has been an adjunct lecturer at the University of South Florida.

Mr. Waggener received a B.S E.E. degree from Rose Hulman Institute of Technology in 1957, an M.E.E. degree from the University of Florida in 1967, and has done postgraduate work at Northeastern University and the University of South Florida. He is a registered professional engineer in Florida and a senior life member of the IEEE.

Index

Abby's sine condition, 103
Acquisition time, 269, 275–77,
      299–302, 307
ACR. *See* Attenuation to the cable loss
Adaptive array, 90–91
ADC. *See* Analog-to-digital converter
Additive white Gaussian noise, 152,
      190–91, 221, 227, 229,
      231, 247–50, 254, 256
ADSL. *See* Asymmetric digital subscriber line
Air-to-air radio link, 52, 87
Airy disk, 105
Aliasing, 22–26, 28–29, 47, 49
Aluminum cable, 124–25
AM-DSB. *See* Amplitude modulation,
      double sideband
AMI code, 156–57, 195, 200
Amplitude modulation, 176–79, 182, 210,
      227–28, 281
Amplitude modulation, double-
      sideband, 195
Amplitude modulation, single sideband, 195
AM-SSB. *See* Amplitude modulation, single
      sideband
Analog-to-analog system, 12–13
Analog-to-digital converter, 19, 29–35
Analog-to-digital system, 12–13
Antennas, 87–90

arrays, 90–93
dielectric lens, 94
horn, 93–94
parabolic dish, 95–96
reflecting, 94–95
Antenna effective noise, 43
Antipodal signals, 193–94, 198, 248
APD. *See* Avalanche photo diode
Asymmetric digital subscriber line, 116
Asymptotic coding gain, 248–49, 255
Atmospheric attenuation, 85, 87, 100–3
Attenuation to the cable loss, 128
Autocorrelation, 36, 38–39, 204
Avalanche photo diode, 140–42, 146
AWGN. *See* Additive white Gaussian noise

B3ZS. *See* Binary three zeros substitution
B6ZS. *See* Binary six zeros substitution
Background noise, 96–98, 107–9
Balun, 129
Bandlimited channel, 23, 27, 180, 182,
      153–54, 159, 188–89, 201–4, 238
Bandpass filter, 228–29, 231, 233, 281
Baseband codes, 152–53
   channel, 164–74
   symbol, 153–64
Baseband detection
   Nyquist symbol, 201–5
   partial response signal, 205–10

Baseband detection (continued)
  rectangular symbol, 198–201
Baseband signaling, 5–7, 151, 177, 194–98
Bayes rule, 292
Beer's equation, 100
Bent pipe satellite link, 7
BER. *See* Bit error rate
Bessel filter, 231
Binary code, 170, 172
Binary *n*-zeros substitution, 167
Binary phase shift keying, 215
Binary six zeros substitution, 157
Binary symmetric channel, 167, 243,
      249–50, 254
Binary three zeros substitution, 157
Biphase code, 153–55, 195
Bipolar Nyquist signal, 212
Bipolar transistor amplifier, 145
Bit energy, 11
Bit error probability, 199–201, 203, 206
Bit error rate, 11–12
Black body, 40
Block code, 167–69, 172, 243,
      246–49, 302–6
BNZS. *See* Binary *n*-zeros substitution
Bolometer, 107
Bose Chaudhuri Hocquenghem
      code, 169–70, 257
Brewster's angle, 133
Broadband communications, 116
Broadside array, 90
BSC. *See* Binary symmetric channel
Butterworth power spectrum, 47–48

Cable attenuation, 128
Cable modem, 129
Cable system, 6, 115–18
      See also Fiber optic system; Wireline
      system
Cable television, 4
Cancellation, 46
Carrier modulation, 280
Carrier recovery, 7
Carrier signaling, 194–98
Carrier synchronization, 269, 280–82
      phase shift keying, 287–96
      suppressed system, 282–87

Carrier-to-cochannel interference ratio, 98
Carrier-to-noise ratio, 103
Cascade network, 41–42
Catadioptric system, 103, 105
CC/I. *See* Carrier-to-cochannel interference
      ratio
CCIR, 85
CCITT, 35
CDM. *See* Code division multiplexing
Center fed dipole, 89
Channel codes
      error correcting, 167–74
      group, 166–67
      randomizing, 164–66
Characteristic impedance, 121, 125, 129
Chromatic dispersion, 135–36
Circular aperture, 104–5
Click noise, 231–33
Clock extraction, 296
Coaxial cable, 115, 125–27, 129, 131–32
Cochannel interference, 97–98
Code division multiplexing, 174–76
Code synchronization, 269–70, 302–5
Code syndrome value, 304–5
Coherent demodulation, 210–26,
      270, 280–81
Coherent detection, 181
Collinear array, 90, 93
Communications link. *See* Transmission
      channel
Companding, 34–35
Complex reflection coefficient, 59
Concatenated code, 167, 171–73, 256
Concatenated Reed-Solomon code, 256–59
Constant envelope, 178, 182
Constraint length, 171
Continuous phase modulation, 185, 220–21
Convolution, 21, 176–77
Convolutional code, 167, 170–71, 190,
      249–56, 302, 305
Copper cable, 115–16, 124–25
Corner reflector, 94
Correlation. *See* Autocorrelation
Correlator word synchronizer, 303–4
Cosine function, 176–77, 203–4, 270
Costas loop, 283–85, 290, 294–95, 298
Coupling, 6, 130, 138–39

CPM. *See* Continuous phase modulation
Cramer-Rao bound, 295
Crosstalk, 128–29, 180, 235
Cutoff filter, 23–24, 27
Cyclic code. *See* Bose Chaudhuri
      Hocquenghem code

Damping factor, 274, 276, 301
DBMSK. *See* Duobinary minimum shift
      keying
Decimation, 27
Decision directed loop, 285–87,
      290–91, 294–95
Decision feedback equalizer, 262–63, 265
Decision threshold, 11–12
Demodulation, 187–88
    coherent, 210–26
    incoherent, 226–34
    simple transmission, 6–7
Density function. *See* Noise probability
      density function
Derandomizer, 165
Design charts
    cable systems, 146–49
    propagation loss, 109–13
    sampling parameters, 49
Detection, 188–98
    fiber optic system, 140–48
    See also Baseband detection
DFE. *See* Decision feedback equalizer
Dielectric lens, 94
Differential delay, 61
Differential phase shift keying, 225–26
Diffraction, 71–73, 75, 78, 99, 105–6
Diffuse multipath, 58
Digital equalizer, 261–63
Digital signal processing, 26–28
Digital subscriber line, 116–17
Digital television transmission, 116–17
Digital-to-analog system, 12–13
Digital-to-digital system, 12–13
Dipole, 89–91, 93
Directivity, 87, 89–90, 92–93
Direct ray, 61, 63, 67
Discrete time model, 235–36, 261, 263
Discrete tone, 180, 227
Discriminator detection, 228–34

Dispersion, 135–36
Distributed transmission line, 119–21
Divergence factor, 66
Doppler effect, 54
Double sideband carrier signal, 212
Double sideband, suppressed carrier, 281–84
Double sideband/phase shift keying, 221–22
DPSK. *See* Differential phase shift keying
Driver, cable, 131–32
DSB-SC. *See* Double sideband, suppressed
      carrier
DSL. *See* Digital subscriber line
DS/PSK. *See* Double sideband/phase shift
      keying
Duobinary minimum shift keying, 220
Duobinary signaling, 161–63, 205–7, 214

Early-late gate synchronizer, 298–99
Edge-emitting light-emitting diode, 138–39
Effective earth radius, 74
Effective noise temperature, 42–43
Egli model, 77
EIA. *See* Electronics Industry Association
Electronics Industry Association, 128
Elgi model, 110
Empirical model, 77–82
Encoding, 151–52
Endfire array, 90
Energy contrast ratio, 10–11, 189
Energy-to-noise-spectral density, 189–90
Envelope detection, 227–28
Environmental noise, 97
EPM-73 model, 78–81, 87, 110–11
Equalization, 259–265
Equalizer training, 307–308
Equivalent noise bandwidth, 39, 48
Error correction, 167–74, 242–46, 302
block code, 246–49
concatenated code, 256–59
convolutional code, 249–56
trellis code modulation, 259
Error performance, 8
Error rate. *See* Bit error rate
Expected value function, 37
Eye pattern, 208–9

Fading, 54, 82, 235, 240–46

Fading (continued)
    short-term, 82–83
Far field, 56, 91, 92
Fast frequency transform, 159
FCC. *See* Federal Communications
       Commission
FDM. *See* Frequency division multiplex
Federal Communications Commission, 77
FFT. *See* Fast frequency transform
Fiber optic system, 115, 129, 131–33
    coupling, 138–39
    design chart, 147–49
    detectors, 140–46
    loss factors, 136
    propagation, 133–37
    transmitters, 139–40
Fiber to the neighborhood, 117
Filtering, 6, 39
    inverse, 260
    nonreturn-to-zero, 238–39
    Nyquist symbol, 156–64
    transverse digital, 261–63
See also Bandpass filter; Lowpass filter;
      Matched filter
Filter/sample detection, 233
Finite impulse response, 28, 163, 207–8,
    235, 265
FIR. *See* Finite impulse response
First-order loop, 273
First-order moment, 36
First-order spectrum, 24–25
Fixed system, 54, 70
Flash converter, 29
Flat earth model, 57–65
Flat panel array, 91
Fokker-Planck equations, 279
Folding process, 22–23
Foliage loss, 85–86
Format synchronization, 269–70, 305–7
Fourier transform, 21, 38, 176–77
Frame synchronization, 303
Free-space loss, 56, 57, 64, 76, 81, 84–85,
    87, 99–100, 110, 111
Free space with multipath, 71–72
Frequency acquisition, 275–77
Frequency dependency, 124

Frequency division multiplexing, 46, 151,
    174–75, 180
Frequency lock, 301
Frequency modulation, 180–81, 198,
    228–34, 241
Frequency reuse, 97
Frequency shift keying, 174, 180–81,
    216–18, 222, 234, 241, 281
Fresnel zone, 71–72
F/S detection. *See* Filter/sample detection
FSK. *See* Frequency shift keying
Full duplex system, 128

Gain, 14, 56, 63–64, 87–89, 93, 112–13,
    141–42, 144, 308
Gaussian noise, 46, 135, 235–37,
    261, 277–80
    signal-to-quantization ratio, 33–34
See also Additive white Gaussian noise
Gaussian probability density
      function, 37, 39, 192
Generation/recombination noise, 108–9
Glint, 58
Golay code, 247
Graded index plastic fiber, 137
G-R noise. *See* Generation/recombination
      noise
Ground ray intercept point, 67
Ground reflection, 58
Ground-to-air radio link, 52
Ground-to-ground radio link, 52, 71, 77
Group code, 166–67, 302

Half Nyquist filter, 159–60
Half rate convolutional code, 171
Hamming distance, 168, 247, 252
Hamming weight, 246
Hard limiting, 182
HDB3. *See* High density bipolar three
HDSL. *See* High rate digital subscriber line
Heaviside differential operator, 271, 288
Hedeman code, 155
Heterodyne detection, 223–26
High density bipolar three, 157
High rate digital subscriber line, 116
High rate unshielded twisted pair cable, 129
Homodyne detection, 223–25, 228

Horizontal polarization, 59, 60, 89
Horn antenna, 93–94
Hybrid optical system, 103

IF. *See* Intermediate frequency
IIR. *See* Infinite impulse response
Image spot diameter, 105
IMD. *See* Intermodulation distortion
Impedance matching, 43–44
Incoherent demodulation, 226–34
Incoherent detection, 181
Infinite impulse response, 28
Infinitesimal dipole, 55–56
Infrared detector, 51, 108
Injection laser diode, 139–40
Input resistance, 145
Input signal format, 12
Integrated Services Digital Network, 116
Interference, 19, 45–49, 96–98, 119, 130,
        154, 157, 161, 188, 190, 201,
        205–6, 235–40, 259–60,
        262, 307–8
Interleaving, 250–51, 257–58
Intermediate frequency, 26, 177, 225, 232
Intermodulation distortion, 46–47
Internet, 115
Inter-Range Instrumentation Group, 153
Intersymbol interference, 119, 154, 157,
        161, 188, 190, 201, 205–6,
        235–40, 259–60, 262, 307–8
Inverse filtering, 260
Inverse square law spreading, 99
IR. *See* Infrared detection
IRIG. *See* Inter-Range Instrumentation
        Group
ISDN. *See* Integrated Services Digital
        Network
Isotropic antenna, 56, 63, 87–89
Iterative coding, 258–59

Jitter, 7–8, 13
Johnson-Gierhart model, 71, 77, 86

Knife-edge obstacle, 73

Lambert power distribution, 138
Laplace variable, 121
Laser diode, 138–40, 155, 178

Last mile problem, 115–16
LED. *See* Light-emitting diode
Light-emitting diode, 99, 101, 104, 106,
        136–40, 155, 178, 228
Lightwave system, 221
Limiter, 287, 291
Linear array, 91–92
Linear filtering, 231, 234–35, 265
Linear sequential shift register, 165
Line of sight, 72–73
LNA. *See* Low noise amplifier
Loading optimum, 34
Lobing effect, 59, 64
Log likelihood ratio, 293–94, 297–98
Lognormal distribution, 82–83
Long distance telecommunication, 4
Longley-Rice model, 53–54, 70–77, 86
Loop filter, 274, 276, 287–88, 290
Lossless mismatch, 43
Lossless transmission line
        propagation, 119–21
Lossy transmission line propagation, 121–23
Low-noise system, 40, 42–44
Lowpass filter, 21, 202, 283, 285, 287
L-RC minimum shift keying, 220

Maksutov catadioptric system, 105
Manchester code. *See* Biphase code
Manmade noise, 96–97
M-ary. *See* Multiphase shift modulation
Matched filter, 191, 194, 198–99, 203,
        208–9, 217–19, 233, 261,
        264, 291, 295, 297
Maximum likelihood decoder, 255
Maximum likelihood estimator, 292–95,
        297, 302–3
Maximum likelihood sequence
        estimation, 263–65
Maximum likelihood synchronizer, 292–94
Maxwell's equations, 55–57
Mean square noise, 45, 144, 202, 204, 219,
        231, 260, 262
Mean square overload error, 33
Mean square phase error, 277–78, 289–91,
        294–96, 302
Mean square quantization error, 31
Mean time to failure, 132

Measurement smoothing formula, 129
Microwave link, 70
Military communications system, 54
Minimum shift keying, 183–85, 218–20
MLE. *See* Maximum likelihood estimator
MLSE. *See* Maximum likelihood sequence
    estimation
MMF. *See* Multiple mode fiber
Mobile system, 54, 70, 81, 97
Modal dispersion, 135–36
Modem, 129
Modulation, 129, 151–52, 174
    amplitude, 176–79, 182, 210, 227–28, 281
    frequency, 180–81, 198, 228–34, 241
    phase, 182–85, 281
    simple transmission model, 5–6
Motion Picture Experts Group, 26
MPEG. *See* Motion Picture Experts Group
MSK. *See* Minimum shift keying
MTBF. *See* Mean time to failure
Multiconductor cable, 129–31
Multilevel signaling, 178, 295
Multimode fiber, 137–38
Multipair cable, 128–29
Multipath, 58, 59, 61–65, 87, 103
Multipath suppression, 84–85
Multipath with diffraction, 71–73
Multiphase shift modulation, 179, 196,
    215–16, 218, 234
Multiple mode fiber, 134–35
Multiplexing, 174–76

NA. *See* Numerical aperture
Near-end crosstalk, 128–29
Near field, 56
Networked system, 1, 4
NEXT. *See* Near-end crosstalk
Node synchronization. *See* Code
    synchronization
Noise, 15, 19, 36–45, 47
    optical system, 107–13, 142–46
    shot, 44–45, 142–45, 148–49
    and system performance, 19
    thermal, 39–44
See also Additive white Gaussian noise;
Gaussian noise; White noise
Noise figure, 15, 40–41, 47

Noise probability density function, 11–12
Noise spectral density, 10, 230
Nonbandlimited channel, 152, 198
Nonbinary code, 170, 172
Nonlinear amplification, 6
Nonlinear quantization, 34
Nonreturn-to-zero, 153–54, 156, 159, 164,
    178, 182, 198–200, 203–4, 209,
    215, 238, 248, 297
Nonstationary noise, 38
Norton equivalent noise, 148–49
NRZ. *See* Nonreturn-to-zero
Numerical aperture, 104, 133–39
Nyquist filter, 207
Nyquist sampling, 20–22
Nyquist signaling, 156–164, 188–89,
    201–5, 209, 211–12

Offset quadrature phase shift
    keying, 182–83, 185
On-off keying, 178, 222–23, 227–28
OOK. *See* On-off keying
Open wire cable, 126
Optical energy, 107
Optical systems
    coherent demodulation, 221–26
    components, 103–6
    coupling, 6
    noise, 107–13
    propagation, 99–103
OQPSK. *See* Offset quadrature phase shift
    keying
Orthogonality, 217
Output frequency spectrum, 5–6
Output signal format, 12
Overload error, 31–34

Pair selected ternary, 157
PAM. *See* Pulse amplitude modulation
Parabolic dish antenna, 94–96, 105, 112–13
Parallel conductor, 125
Parametric amplifier, 43
Partial response signaling, 159, 161–64,
    205–10, 214, 297
Path absorption, 99
Path loss, 67–68, 71, 98–99
PCM. *See* Pulse code modulation

PCS. *See* Personal communication system
PDM. *See* Pulse duration modulation
Performance, system, 3–4, 6
Personal communication system, 86, 97
Phase acquisition, 275
Phase locked loop, 270–81, 283, 285, 297, 299–300
  first order, 273
  second order, 274, 276, 278
Phase modulation, 182–85, 281
Phase shift, 58, 60–63
Phase shift keying, 17, 174, 178, 182–185, 196, 212, 215–16, 220–23, 225, 239, 248, 282, 293
Phase shift key synchronization, 287–96
Photoconductor, 107–9
Photodetector, 44, 107–8
Photodiode, 140–46, 223–25
Photoemissive detector, 107, 109
Photomultiplier, 109
Photon arrival rate, 107–8
Photon-limited noise, 142–45
Photovoltaic detector, 107, 109
Pilot tone, 211
Plain old telephone service, 128
Plane earth model, 86, 110
Plastic optical fiber, 136–137
PLL. *See* Phase locked loop
PMMA. *See* Polymethyl methacrylate
POF. *See* Plastic optical fiber
Point-to-point system, 1–4
Poisson density function, 45, 222
Poisson photo current, 142
Polymethyl methacrylate, 137
POTS. *See* Plain old telephone service
PPM. *See* Pulse position modulation
Preamplifier noise, 142–46
Precoded duobinary, 207
Precoding, 208
Prefilter frequency response, 61, 162–63
Probability density function, 36–37, 279
Propagation, 53–54
  Egli model, 77
  EPM-73 model, 78–81
  fiber optic cable, 133–37
  flat earth model, 57–65
  fundamentals, 55–57

Longley-Rice model, 70–77
  model choice, 86–87
  optical system, 99–103
  other losses, 84–86
  spherical earth model, 65–70
  urban models, 81–82
  variability, 82–84
  wireline system, 118–25
PR sequence. *See* Pseudo random sequence
Pseudo random sequence, 165–66, 302
PSK. *See* Phase shift keying
PST. *See* Pair selected ternary
Pull-out frequency, 276
Pulse amplitude modulation, 156, 158, 178, 195–96, 199, 281
Pulse code modulation, defined, 1
Pulse duration modulation, 156, 158
Pulse position modulation, 156, 158

QAM. *See* Quadrature amplitude modulation
QFSK. *See* Quadrature frequency shift keying
QPSK. *See* Quadrature phase shift keying
Quadrature amplitude modulation, 173, 178–79, 182, 195–97, 211–14, 259, 281
Quadrature frequency shift keying, 218
Quadrature phase shift keying, 178, 182, 185, 196, 212, 215, 221, 259, 295
Quantization, 19, 29–35, 256
Quantum detection limit, 228
Quantum efficiency, 108, 140–41
Quantum nature of light, 107, 222

Radio frequency propagation. *See* Propagation
Radio frequency system, 51–53
Radio horizon, 67, 70
RADSL. *See* Rate adaptive digital subscriber line
Raised cosine detection, 203–4, 297
Randomizing code, 164–66
Random site system, 70–71
Rate adaptive digital subscriber line, 116
Ray divergence, 66
Rayleigh fading, 82, 240–44

Rayleigh's law, 102
Ray tracing, 54, 65, 86, 133
Receiver, 6, 107, 131–32
Reciprocity, 6, 88
Rectangular symbol, 153–56, 198–201
Recursive systematic convolutional code, 258
Redundancy, data stream, 168
Reed-Solomon code, 170, 172, 256
Reflecting antenna, 94–96
Reflection coefficient, 59, 60–64, 70
Reflective optical system, 103, 105
Refraction index, 65, 73–74, 133–36
Refractive corrector lens, 105
Refractive optical system, 103, 105
Repeater, 7, 132
Reset integrator, 231, 233
Residual timing error, 269
Resistance matching, 40
Return-to-zero, 156, 178
RF system. *See* Radio frequency system
Rician fading, 82
Rising noise, 232
RMS. *See* Root mean square
Root mean square, 12–13, 25, 45, 47, 49,
        145, 192
Roughness factor, 74–75, 77, 87
RS-232 standard, 126
RS-422/423 standard, 126–27
RSC. *See* Recursive systematic convolutional
        code
RS code. *See* Reed-Solomon code
RSI. *See* Reset integrator
RZ. *See* Return-to-zero

Sample-whitened matched filter, 261
Sampling
    aliasing, 22–26
    Nyquist, 29–32
    sample rate reduction, 26–29
Satellite communications, 4, 7, 46
Satellite television link, 115–16
Satellite-to-satellite link, 52, 99
Scattering loss, 58, 100, 102–3
Schmidt catadioptric system, 105
Second order loop, 274, 276, 278
Second order moment, 37
Semiconductor laser, 223, 228

Sequence transition, 164–67
Shannon bound, 189–90
Shannon code, 173
Shielded twisted pair, 128
Short dipole, 89–90
Shot noise, 44–45, 109, 142–45, 148–49
Shunt capacitance, 145
Signal-to-noise ratio, 6, 9–11, 15–16,
        41–43, 107, 109, 142, 145–46,
        165, 180, 189–90, 194, 203–4,
        206, 242, 245–46, 258–59,
        278–80, 298, 302
Signal-to-quantizing ratio, 31–34, 48
Sine function, 176–77
Sinewave signal-to-quantization
        ratio, 31–33, 48
Single mode fiber, 134–35, 137–38
Single sideband, 282, 283
Single sideband amplitude modulation, 211
Skin resistance, 124
Sliding block decoder, 250
Sliding correlator, 303
Snell's law, 58, 133
Soft decision decoding gain, 248
Source compression, 19
Source encoder, 19
Specular ray, 58
Specular reflection, 64
Spherical earth model, 65–70, 87
Spread spectrum, 221
Spurious signal, 6
SQR. *See* Signal-to-quantizing ratio
Square root filter, 159–60
Squaring loop, 283, 288–90, 295
SSB. *See* Single sideband
SSB/AM. *See* Single sideband amplitude
        modulation
Standard deviation model, 83–84
Standard earth atmosphere, 65
Standardization, radio system, 151–52
Stationary noise, 38
Step index fiber, 135, 137
STP. *See* Shielded twisted pair
Subrate sampling, 27–29
Suburban model, 83, 86
Suppressed carrier, 281–87

Surface-emitting light-emitting
    diode, 138–39
Symbol codes, 129
    Nyquist, 156–64
    rectangular, 153–56
Symbol decisions, 189, 206–8, 235
Symbol energy-to-noise spectral
    density, 152–53
Symbol synchronization, 269, 287, 296–302
Synchronization, 8–9, 13–14, 269–80
See also Carrier synchronization; Symbol
    synchronization
Synchronization threshold, 269
Syndrome detection, 249–51, 304–5
Systematic code, 249–51, 255, 259, 305

Tamed frequency modulation, 220
Tap gain, 308
TCM. *See* Trellis code modulation
TDM. *See* Time division multiplexing
Telecommunications Industry
    Association, 128
Telecommunications system, 115–16
Terrain profile, 74–75, 77, 87
TFM. *See* Tamed frequency modulation
Thermal detector, 107
Thermal noise, 39–44
Third order product, 46
TIA. *See* Telecommunications Industry
    Association
Time dispersion, 135–36
Time division multiplexing, 174–75
Time for synchronization, 8–9, 13
Time to lose lock, 269, 306–7
Timing error, 302–3
Timing jitter, 7–8, 13
Timing margin, 208–9
Tone detection, 228
Transformer coupling, 130
Transimpedance, 145
Transmission channel, 14
    design, 14–17
    performance, 3–4
Transmission models
    simple, 3, 5–7
    with transponder, 4, 7–9
Transmitters

    fiber optic system, 139–40
    simple model, 6
Transponder, 4, 7–9
Transversal equalizer, 261
Transverse digital equalizer, 261–63
Traveling wave tube, 46
Trellis code modulation, 168, 173, 259
Turbo code, 168, 173–74, 256, 257–59
Twinaxial cable, 129–30
Twisted-pair cable, 126–29
Two-ray model, 75
TWT. *See* Traveling wave tube

UAV. *See* Unmanned airborne vehicle
UHF. *See* Ultra high frequency
Ultra high frequency, 64, 77, 97
Union bound, 248, 254
Unipolar code, 155, 178
Unmanned airborne vehicle, 17
Unmanned earth vehicle, 69
Unshielded twisted pair, 128–29
Upper normal distribution, 192
Urban noise, 81–82, 86, 96, 110–12
UTP. *See* Unshielded twisted pair

VA. *See* Viterbi algorithm
VCO. *See* Voltage controlled oscillator
Vector representation, 193
Vertical polarization, 59–62, 89
Very high frequency, 52, 77, 83, 97
Vestigial sideband, 282
VF. *See* Voltage to frequency
VHF. *See* Very high frequency
Video compression, 26
Video signal sampling, 25–26
Viterbi algorithm, 255–256, 263–65
Viterbi decoder, 190, 249, 251–55, 263, 305
Voltage controlled oscillator, 270–76,
    283–85, 287
Voltage divider network, 108
Voltage standing wave ratio, 44
Voltage to frequency, 29
VSB. *See* Vestigial sideband
VSWR. *See* Voltage standing wave ratio

Weather, 54, 84, 87
Weighting factor, 75
Whitening filter, 261

White noise, 10, 39, 45–46,, 152, 190–91,
        221, 227, 229, 231, 247–50,
        254, 256
Wideband services, 115–16, 129
Wireless systems, 115

Wireline systems, 115–18
    cable characteristics, 125–31
    cable drivers/receivers, 131
    propagation, 118–25

The Artech House Telecommunications Library

Vinton G. Cerf, Series Editor

Access Networks: Technology and V5 Interfacing, Alex Gillespie

Advanced High-Frequency Radio Communications,
  Eric E. Johnson, Robert I. Desourdis, Jr., et al.

Advanced Technology for Road Transport: IVHS and ATT,
  Ian Catling, editor

Advances in Computer Systems Security, Vol. 3, Rein Turn, editor

Advances in Telecommunications Networks, William S. Lee and
  Derrick C. Brown

*Advances in Transport Network Technologies: Photonics
  Networks, ATM, and SDH,* Ken-ichi Sato

An Introduction to International Telecommunications Law,
  Charles H. Kennedy and M. Veronica Pastor

*Asynchronous Transfer Mode Networks: Performance Issues,
  Second Edition,* Raif O. Onvural

ATM Switches, Edwin R. Coover

ATM Switching Systems, Thomas M. Chen and Stephen S. Liu

Broadband: Business Services, Technologies, and Strategic Impact,
  David Wright

Broadband Network Analysis and Design, Daniel Minoli

Broadband Telecommunications Technology, Byeong Lee,
  Minho Kang and Jonghee Lee

Cellular Mobile Systems Engineering, Saleh Faruque

Cellular Radio: Analog and Digital Systems, Asha Mehrotra

Cellular Radio: Performance Engineering, Asha Mehrotra

Cellular Radio Systems, D. M. Balston and R. C. V. Macario, editors

CDMA for Wireless Personal Communications, Ramjee Prasad

Client/Server Computing: Architecture, Applications, and Distributed Systems Management, Bruce Elbert and Bobby Martyna

Communication and Computing for Distributed Multimedia Systems, Guojun Lu

Communications Technology Guide for Business, Richard Downey, et al.

Community Networks: Lessons from Blacksburg, Virginia, Andrew Cohill and Andrea Kavanaugh, editors

Computer Networks: Architecture, Protocols, and Software, John Y. Hsu

Computer Mediated Communications: Multimedia Applications, Rob Walters

Computer Telephone Integration, Rob Walters

Convolutional Coding: Fundamentals and Applications, Charles Lee

Corporate Networks: The Strategic Use of Telecommunications, Thomas Valovic

The Definitive Guide to Business Resumption Planning, Leo A. Wrobel

Desktop Encyclopedia of the Internet, Nathan J. Muller

Digital Beamforming in Wireless Communications, John Litva and Titus Kwok-Yeung Lo

Digital Cellular Radio, George Calhoun

Digital Hardware Testing: Transistor-Level Fault Modeling and Testing, Rochit Rajsuman, editor

Digital Switching Control Architectures, Giuseppe Fantauzzi

Digital Video Communications, Martyn J. Riley and Iain E. G. Richardson

Distributed Multimedia Through Broadband Communications Services, Daniel Minoli and Robert Keinath

Distance Learning Technology and Applications, Daniel Minoli

EDI Security, Control, and Audit, Albert J. Marcella and Sally Chen

Electronic Mail, Jacob Palme

Enterprise Networking: Fractional T1 to SONET, Frame Relay to BISDN, Daniel Minoli

Expert Systems Applications in Integrated Network Management, E. C. Ericson, L. T. Ericson, and D. Minoli, editors

FAX: Digital Facsimile Technology and Applications, Second Edition, Dennis Bodson, Kenneth McConnell, and Richard Schaphorst

FDDI and FDDI-II: Architecture, Protocols, and Performance, Bernhard Albert and Anura P. Jayasumana

Fiber Network Service Survivability, Tsong-Ho Wu

Future Codes: Essays in Advanced Computer Technology and the Law, Curtis E. A. Karnow

Guide to Telecommunications Transmission Systems, Anton A. Huurdeman

A Guide to the TCP/IP Protocol Suite, Floyd Wilder

Implementing EDI, Mike Hendry

Implementing X.400 and X.500: The PP and QUIPU Systems, Steve Kille

Inbound Call Centers: Design, Implementation, and Management, Robert A. Gable

Information Superhighways Revisited: The Economics of Multimedia, Bruce Egan

Integrated Broadband Networks, Amit Bhargava

International Telecommunications Management, Bruce R. Elbert

International Telecommunication Standards Organizations,
  Andrew Macpherson

Internetworking LANs: Operation, Design, and Management,
  Robert Davidson and Nathan Muller

Introduction to Document Image Processing Techniques,
  Ronald G. Matteson

Introduction to Error-Correcting Codes, Michael Purser

An Introduction to GSM, Siegmund Redl, Matthias K. Weber and
  Malcom W. Oliphant

*Introduction to Radio Propagation for Fixed and Mobile
  Communications,* John Doble

Introduction to Satellite Communication, Second Edition,
  Bruce R. Elbert

Introduction to T1/T3 Networking, Regis J. (Bud) Bates

Introduction to Telecommunications Network Engineering,
  Tarmo Anttalainen

*Introduction to Telephones and Telephone Systems, Second
  Edition,* A. Michael Noll

Introduction to X.400, Cemil Betanov

LAN, ATM, and LAN Emulation Technologies, Daniel Minoli and
  Anthony Alles

Land-Mobile Radio System Engineering, Garry C. Hess

LAN/WAN Optimization Techniques, Harrell Van Norman

LANs to WANs: Network Management in the 1990s,
  Nathan J. Muller and Robert P. Davidson

The Law and Regulation of Telecommunications Carriers,
  Henk Brands and Evan T. Leo

Minimum Risk Strategy for Acquiring Communications Equipment and Services, Nathan J. Muller

Mobile Antenna Systems Handbook, Kyohei Fujimoto and J. R. James, editors

Mobile Communications in the U.S. and Europe: Regulation, Technology, and Markets, Michael Paetsch

Mobile Data Communications Systems, Peter Wong and David Britland

Mobile Information Systems, John Walker

Networking Strategies for Information Technology, Bruce Elbert

Packet Switching Evolution from Narrowband to Broadband ISDN, M. Smouts

Packet Video: Modeling and Signal Processing, Naohisa Ohta

Performance Evaluation of Communication Networks, Gary N. Higginbottom

Personal Communication Networks: Practical Implementation, Alan Hadden

Personal Communication Systems and Technologies, John Gardiner and Barry West, editors

Practical Computer Network Security, Mike Hendry

Principles of Secure Communication Systems, Second Edition, Don J. Torrieri

Principles of Signaling for Cell Relay and Frame Relay, Daniel Minoli and George Dobrowski

Principles of Signals and Systems: Deterministic Signals, B. Picinbono

Private Telecommunication Networks, Bruce Elbert

Pulse Code Modulation Systems Design, William N. Waggener

Radio-Relay Systems, Anton A. Huurdeman

RF and Microwave Circuit Design for Wireless Communications,
Lawrence E. Larson

The Satellite Communication Applications Handbook,
Bruce R. Elbert

Secure Data Networking, Michael Purser

Service Management in Computing and Telecommunications,
Richard Hallows

Signaling in ATM Networks, Raif O. Onvural, Rao Cherukuri

Smart Cards, José Manuel Otón and José Luis Zoreda

Smart Card Security and Applications, Mike Hendry

Smart Highways, Smart Cars, Richard Whelan

SNMP-Based ATM Network Management, Heng Pan

*Successful Business Strategies Using Telecommunications
Services,* Martin F. Bartholomew

Super-High-Definition Images: Beyond HDTV, Naohisa Ohta,
et al.

Telecommunications Deregulation, James Shaw

Television Technology: Fundamentals and Future Prospects,
A. Michael Noll

Telecommunications Technology Handbook, Daniel Minoli

Telecommuting, Osman Eldib and Daniel Minoli

Telemetry Systems Design, Frank Carden

Teletraffic Technologies in ATM Networks, Hiroshi Saito

*Toll-Free Services: A Complete Guide to Design, Implementation,
and Management,* Robert A. Gable

Transmission Networking: SONET and the SDH, Mike Sexton and
Andy Reid

Troposcatter Radio Links, G. Roda

Understanding Emerging Network Services, Pricing, and Regulation, Leo A. Wrobel and Eddie M. Pope

Understanding GPS: Principles and Applications, Elliot D. Kaplan, editor

Understanding Networking Technology: Concepts, Terms and Trends, Mark Norris

UNIX Internetworking, Second Edition, Uday O. Pabrai

Videoconferencing and Videotelephony: Technology and Standards, Richard Schaphorst

Voice Recognition, Richard L. Klevans and Robert D. Rodman

Wireless Access and the Local Telephone Network, George Calhoun

Wireless Communications in Developing Countries: Cellular and Satellite Systems, Rachael E. Schwartz

Wireless Communications for Intelligent Transportation Systems, Scott D. Elliot and Daniel J. Dailey

Wireless Data Networking, Nathan J. Muller

Wireless LAN Systems, A. Santamaría and F. J. López-Hernández

Wireless: The Revolution in Personal Telecommunications, Ira Brodsky

World-Class Telecommunications Service Development, Ellen P. Ward

Writing Disaster Recovery Plans for Telecommunications Networks and LANs, Leo A. Wrobel

X Window System User's Guide, Uday O. Pabrai

For further information on these and other Artech House titles, including previously considered out-of-print books now available through our In-Print-Forever™ (IPF™) program, contact:

Artech House
685 Canton Street
Norwood, MA 02062
781-769-9750
Fax: 781-769-6334
Telex: 951-659
email: artech@artech-house.com

Artech House
Portland House, Stag Place
London SW1E 5XA England
+44 (0) 171-973-8077
Fax: +44 (0) 171-630-0166
Telex: 951-659
email: artech-uk@artech-house.com

Find us on the World Wide Web at:
www.artech-house.com